Dr. Chandan Deep Singh

Electrical Discharge Machining

Optimization of chromium powder mixed EDM parameters during machining of H13 tool steel

Anchor Academic Publishing

Singh, Chandan Deep: Electrical Discharge Machining. Optimization of chromium powder mixed EDM parameters during machining of H13 tool steel, Hamburg, Anchor Academic Publishing 2018

Buch-ISBN: 978-3-96067-210-4
PDF-eBook-ISBN: 978-3-96067-710-9
Druck/Herstellung: Anchor Academic Publishing, Hamburg, 2018

Bibliografische Information der Deutschen Nationalbibliothek:
Die Deutsche Nationalbibliothek verzeichnet diese Publikation in der Deutschen Nationalbibliografie; detaillierte bibliografische Daten sind im Internet über http://dnb.d-nb.de abrufbar.

Bibliographical Information of the German National Library:
The German National Library lists this publication in the German National Bibliography. Detailed bibliographic data can be found at: http://dnb.d-nb.de

All rights reserved. This publication may not be reproduced, stored in a retrieval system or transmitted, in any form or by any means, electronic, mechanical, photocopying, recording or otherwise, without the prior permission of the publishers.

Das Werk einschließlich aller seiner Teile ist urheberrechtlich geschützt. Jede Verwertung außerhalb der Grenzen des Urheberrechtsgesetzes ist ohne Zustimmung des Verlages unzulässig und strafbar. Dies gilt insbesondere für Vervielfältigungen, Übersetzungen, Mikroverfilmungen und die Einspeicherung und Bearbeitung in elektronischen Systemen.

Die Wiedergabe von Gebrauchsnamen, Handelsnamen, Warenbezeichnungen usw. in diesem Werk berechtigt auch ohne besondere Kennzeichnung nicht zu der Annahme, dass solche Namen im Sinne der Warenzeichen- und Markenschutz-Gesetzgebung als frei zu betrachten wären und daher von jedermann benutzt werden dürften.

Die Informationen in diesem Werk wurden mit Sorgfalt erarbeitet. Dennoch können Fehler nicht vollständig ausgeschlossen werden und die Diplomica Verlag GmbH, die Autoren oder Übersetzer übernehmen keine juristische Verantwortung oder irgendeine Haftung für evtl. verbliebene fehlerhafte Angaben und deren Folgen.

Alle Rechte vorbehalten

© Anchor Academic Publishing, Imprint der Diplomica Verlag GmbH
Hermannstal 119k, 22119 Hamburg
http://www.diplomica-verlag.de, Hamburg 2018
Printed in Germany

ABSTRACT

In the present study, optimization of chromium powder mixed EDM parameters is studied during machining of H13 tool steel. Four input parameters of powder mixed EDM namely peak current, pulse on time, duty cycle and powder concentration are varied, each at three levels, to get the optimum responses. Material removal rate (MRR), Tool wear rate (TWR) and Surface Roughness (Ra) are considered as performance measures. Copper electrode of 16 mm is used as the tool. Response Surface Methodology is used to correlate input and output parameters. The variation of responses due to variation in input parameters has been studied and shown in the form of surface plots and contour plots. Results reveals that pulse on time, powder concentration, duty cycle and peak current are the significant factors affecting MRR while TWR is significantly affected by peak current only. In case of Surface Roughness, pulse on time, peak current and powder concentration are the significant factors affecting the Ra. Furthermore, it is found that maximum MRR is there when both peak current and duty cycle are increased simultaneously. Minimum TWR is obtained when current is at its low level (10 Amp). In case of surface roughness, minimum value of Ra is obtained when pulse on time and peak current simultaneously are at low levels (100μsec and 10 amp) while powder concentration and duty cycle are at their intermediate levels (10g/l and 6%). Desirability Method is used to optimize the input parameters to get optimal values of responses. Optimal solution that has been found for the present study is 16.1250 mm^3/min MRR, 0.3161 mm^3/min TWR and 7.6987 μm Ra with peak current 20 ampere, duty cycle 5.0909%, pulse on time 200 μsec and powder concentration of 11.667 g/l.

Keywords: MRR, TWR, Surface Roughness (Ra), RSM and Desirability Method.

TABLE OF CONTENTS

Chapter-1 Introduction	1-14
1.1 Introduction to Non Conventional Machining Processes	1
1.2 Electrical Discharge Machining	2
1.3 Working Principle of EDM	3
1.4 Mechanism of Material Removal in EDM	4
1.5 Powder Mixed EDM	6
1.6 Major Components of EDM or PMEDM	7
1.7 Important Process Parameters of PMEDM or EDM	9
1.8 General Requirements of Dielectric Medium in EDM	11
1.9 Flushing	11
1.10 Applications of EDM	13
1.11 Advantages of EDM	13
1.12 Disadvantages of EDM	14
Chapter-2 Literature Review & Problem Formulation	15-27
2.1 Literature Review	15
2.2 Problem Formulation	25
2.3 Objectives of the Present Study	27
Chapter-3 Optimization Technique Used	28-32
3.1 Response Surface Methodology	28
3.2 Surface Plot	29
3.3 Contour Plot	30
3.4 Desirability Approach	31
3.5 Optimization Plot by Desirability Method	32

Chapter-4 Experimental Setup   **33-38**

4.1 EDM Machine   33

4.2 Experimental Setup for Powder Mixed in the Dielectric   34

4.3 Preparing Workpiece for Experimentation   36

4.4 Weighing the Specimen   37

4.5 Specification of Balance   38

4.6 Surface Roughness Measurement   38

Chapter-5 Results & Discussion   **39-68**

5.1 Response Surface Methodology   39

5.2 Input Parameters and their Levels   40

5.3 Responses Variables   41

5.4 Design Matrix and Observation Table   43

5.5 Results for Material Removal Rate   46

5.6 Results for Tool Wear Rate   52

5.7 Results for Surface Roughness (Ra)   57

5.8 Optimization of the Powder Mixed EDM Parameters   63

Chapter-6 Conclusion & Future Scope   **69-71**

6.1 Conclusion   69

6.2 Future Scope   70

References   **72-77**

LIST OF FIGURES

Fig. No.	Title	Page No.
1.1	Setup of EDM	3
1.2	Working principle of EDM	3
1.3	Diagram of the EDM physical process	5
1.4	Working Principle of PMEDM	6
1.5	Positioning System	7
1.6	Tool Holder	8
1.7	Concept of pulse on time and pulse off time	10
1.8	Normal and Reverse Polarity	11
1.9	Injection flushing through electrode	12
1.10	Side flushing	12
1.11	Suction flushing through electrode	12
3.1	Surface plot generated by RSM	30
3.2	Contour plot generated by RSM	30
3.3	Optimization plot by Desirability Method	32
4.1	EDM used for experimentation	33
4.2	Experimental set up for powder mixed dielectric EDM	35
4.3	Specimens for experimentation	36
4.4	Electronic balance	37
5.1	Design of experiment by RSM using MINITAB software	39
5.2 (a)	Workpiece after machining	42
5.2 (b)	Workpiece before machining	42
5.3	Copper electrode used for machining	42

Fig. No.	Title	Page No.
5.4	Main effects plot for MRR	49
5.5	Surface plot and Contour plot for MRR	50
5.6	Residual Plots for MRR	51
5.7	Main effects plot for TWR	54
5.8	Surface plots for TWR	55
5.9	Contour plot for TWR	56
5.10	Residual plots for TWR	57
5.11	Main effects plot for Ra	59
5.12	Surface plot for Surface roughness (Ra)	60
5.13	Contour plots for Surface Roughness (Ra)	61
5.14	Residual plots for Surface roughness (Ra)	62
5.15	Optimal plot for MRR	64
5.16	Optimal plot for TWR	65
5.17	Optimal plot for Surface roughness (Ra)	66
5.18	Optimization plot for the experiment	68

LIST OF TABLES

Table No.	Title	Page No.
4.1	Specification of EDM machine used for experimentation	34
4.2	Composition of H13 die steel	37
4.3	Specification of Electronic balance	38
5.1	Input Parameters with Levels	40
5.2	Constant Parameters	41
5.3	Design matrix for experimentation	43
5.4	Response values after EDM operation	45
5.5	ANOVA table for MRR	47
5.6	ANOVA table for TWR	52
5.7	ANOVA table for Ra	58
5.8	Single objective optimization table for MRR	63
5.9	Single objective optimization table for TWR	64
5.10	Single objective optimization table for Ra	65
5.11	Validation table for Desirability Method	66
5.12	Multi-objective optimization by Desirability Method	67

List of Abbreviations & Symbols

EDM = Electric Discharge Machining

PMEDM = Powder Mixed Electric Discharge Machining

MRR = Material removal Rate

TWR = Tool Wear Rate

Ra = Average Surface Roughness

RSM = Response Surface Methodology

ANOVA = Analysis of Variance

$\mu sec = 10^{-6}$ sec

$\mu m = 10^{-6}$ m

t = Time for Machining (8 min)

Wtb = Weight of tool before machining in grams

Wta = Weight of tool after machining in grams

ρ = Density of copper = $8.96 gm/cm^3$

Wjb = Weight of the workpiece before machining in grams

Wja = Weight of the workpiece after machining in grams

$\rho,$ = Density of H13 steel = $7.80 gm/cm^3$

CHAPTER 1

INTRODUCTION

1.1 Introduction to Non Conventional Machining Processes

Traditional or Conventional machining processes work on the principle that there should be the physical contact between the tool and the workpiece and the tool must be harder than the workpiece for the removal of the material. But newly developed materials such as carbides, nickel based alloys, Hastelloy, Inconel, hot die steels etc have very high strength, hardness, corrosion resistance and other properties which make them almost impossible to machine with conventional processes. Therefore, there is a need to develop new tool materials and processes for the machining of such type of materials with high accuracy and productivity. Non conventional processes provide solution for machining of these types of advanced materials. These processes are non conventional in the sense that these do not use the tool to remove the material from the workpiece but use the energy for material removal hence material is removed without the formation of chips.

Based on the type of energy used by these processes for the machining of workpiece, these are classified into following type of categories.

a) **Mechanical processes:** These are the processes in which the material is removed from the workpiece by the mechanical action of abrasive particle or fluid or both. In some cases mechanical action is achieved by vibrating the abrasive particles at high frequency as in Ultrasonic machining, while in Abrasive jet machining and Water jet machining, kinetic energy of abrasive jet and fluid respectively striking the workpiece provides the mechanical action for the erosion of the material.

b) **Electrochemical processes:** These processes require electrochemical energy for the removal of metal from the workpiece. Material is removed from the anode workpiece and transported to the cathode tool in an electrolyte bath. Electrolyte flows rapidly between the two poles to prevent the plating of the eroded material on to the tool. Electrochemical machining, electrochemical grinding and electrochemical deburring are the processes that utilize electrochemical energy for material removal.

c) **Chemical processes:** It involves the application of strong chemical etchant to remove the material from desired portion of the workpiece. The remaining portion of the workpiece is covered with the maskant to prevent the etching from this undesired portion. Chemical milling, chemical blanking and photochemical machining are the examples of this process.

d) **Thermal processes:** These processes utilize the energy in the form of localized heat, light, electron bombardment for material removal by melting or vaporizing the area of the workpiece from where material is to be removed. Examples are electron beam machining, laser beam machining, plasma arc machining and electric discharge machining.

The selection of processes depends on various factors like process parameters, process capabilities, shapes to be machined, properties of the material and economics of the process.

1.2 Electrical Discharge Machining (EDM)

EDM is one of the extensively used non conventional machining processes for material removal. In this, material is removed by the initiation of electrical discharge between the tool and the workpiece. There is no direct contact between tool and workpiece. It is capable of machining electrical conductive materials regardless their hardness and widely used in aerospace industry, automobiles industry, die and mould making industry to machine hard materials and their alloys. General set up of electrical discharge machine is shown in fig 1.1

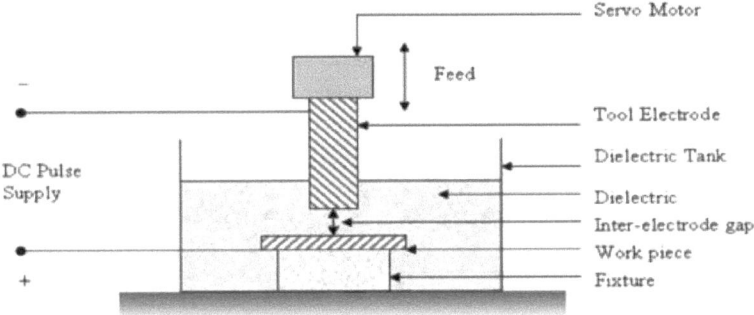

Fig 1.1 Setup of EDM [35]

1.3 Working Principle of EDM

EDM is a thermal process that uses spark discharges to machine the material. A shaped electrode acts as a tool which makes cavities or holes in the workpiece. Electrically conductive workpiece is connected to one pole of a pulsed power supply and electrode or tool is connected to another pole of power supply.

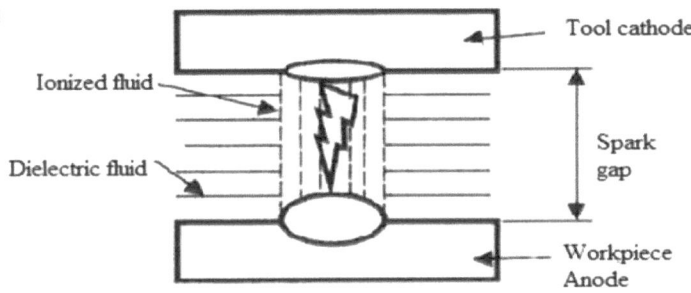

Fig 1.2 Working principle of EDM [35]

A small gap is maintained between the electrode and the workpiece to provide electrical resistance in gap. An intensive electric field is created between the tool and the workpiece when a pulse of D.C electricity is delivered. This electric field is created at a point where surface irregularities provide the narrowest gap. As a result of this field, naturally occurring microscopic contaminants suspended in the dielectric fluid and the negatively charged particles emitted from the workpiece form high conductive bridge across the gap.

As the voltage increases in the beginning, the temperature of this bridge increases and formation of spark comes into play between two surfaces. At the mid- point of electrical pulse, voltage is decreased by power supply and current is increased. Due to this, increase in temperature and pressure in the spark channel takes place. Because of this increase in temperature and pressure, small amount of material melts and vaporizes from both electrode and workpiece at the point of spark contact Fed by gaseous by products of vaporization, a bubble rapidly expand outward from the spark channel. The spark and heating are stopped when the electrical pulse is terminated. This causes both the spark channel and the vapour bubble to collapse. The injection of relatively cool dielectric fluid results in an explosive expulsion of molten metal from both surfaces, resulting in the formation of small crater in both surfaces. The entire sequence takes place in only micro-seconds to mini-seconds. As there is no contact between the tool and the workpiece, so there is no force generated during machining.

1.4 Mechanism of Material Removal in EDM

In EDM, material is removed from the workpiece and the electrode by the series of sparks at the closest point which decrease the distance between the electrode and the workpiece. The next spark occurs at the next closest point. Material at the closest point is heated. This results in vaporization of material due to origination and termination of spark. Whole sequence of material removal mechanism is discussed below with the help of fig 1.3.

In fig 1.3(a) when the voltage is applied between the tool and the workpiece, a strong electric field is developed at the point of least distance between the tool and the workpiece. Increasing the voltage ionizes the dielectric fluid. Fig 1.3(b) indicates that when the voltage reached its peak value insulating properties of the dielectric decreases along the narrow channel centered in the strongest part of the field. A current is established as the number of ionic particles increase and discharge channel is formed.

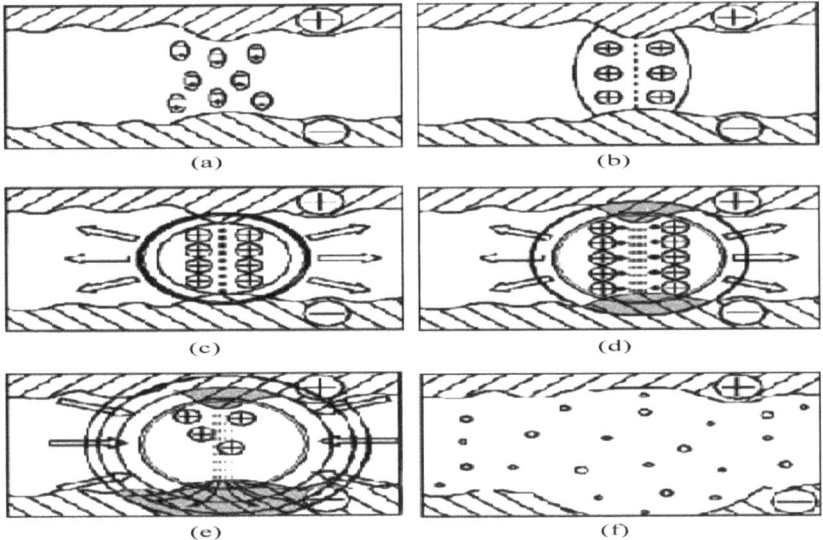

Fig 1.3 Diagram of the EDM physical process [36]

In fig 1.3(c), due to the buildup of heat with increase in current in the discharge channel, a plasma channel is formed consists of vaporized material from the electrode and the workpiece. At the end of the pulse on time, the pressure inside the plasma is nearly at its peak point which causes the growth of vapour bubbles as shown in fig 1.3(d). When the pulse voltage ceases, the plasma channel collapses as there is sharp decrease in the plasma channel pressure and the molten cavities explode violently into the dielectric liquid as represented in fig 1.3(e). Finally the surface cools down instantaneously, where all vaporized and fraction of melted material in the form of irregularly shaped or hollow spherical particles are flushed away by the dielectric liquid.

As there is no direct contact between the tool and the workpiece, therefore problems like vibration, chattering and mechanical stresses does not occur in EDM during machining of components. In spite of these advantages, problems like poor surface finish, low material removal rate etc prevent its uses in some industries to some extent. In order to remove these problems, a new technology known as powder mixed electric discharge machining (PMEDM) has been emerged to improve the machining performances of conventional EDM

1.5 Powder Mixed EDM

In PMEDM, a suitable powder like silicon, vanadium, titanium etc is mixed with the dielectric of EDM. Due to the applied electric field applied, the powder particles get energized and accelerated and become conductors and promote breakdown in the gap and also enhance the spark gap between the tool and the workpiece.

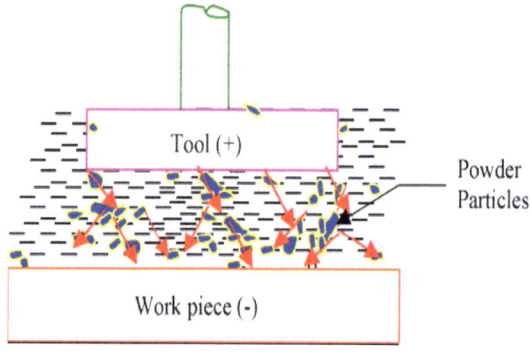

Fig 1.4 Working Principle of PMEDM [30]

Powder particles formed the chain type structure and arrange themselves in the direction of current. This causes the bridging gap between the electrode and workpiece, hence insulating strength of dielectric gets reduced which lead to easy short circuiting and hence early explosion in the gap takes place. It results in the series of discharges under the electrode area. Due to this, faster sparking causes the faster erosion from the workpiece and hence MRR get increased. Addition of powder in the dielectric enlarged the plasma which causes the electric density to decrease and hence uniform erosion occurs on the workpiece leading to better surface finish along with high MRR.

1.6 Major Components of EDM or PMEDM

   a) Positioning System
   b) Servo Feed System
   c) Power Supply System
   d) Tool Holder
   e) Dielectric System
   f) Machining Tank

These systems are discussed in brief as following:

   a) Positioning System: It consists of CNC two axes table. This system is used to provide the movement to the workpiece in X and Y direction.

Fig 1.5 Positioning System

   b) Servo Feed System: This system maintains the constant gap between the tool and the workpiece throughout the machining operation by sensing and comparing the gap voltage with the present value. If any difference exists, then this difference is used to control the movement of servomotor to adjust the gap.

c) **Power Supply System:** Power supply system converts AC from main supply into pulse DC required to produce spark discharges. Firstly, the input power is transformed into continuous DC power by solid state rectifiers. A square wave signal via digital multi vibrator oscillator circuit is generated by using small percentage of DC power supply. This signal is used to start power transistors. These transistors act as high speed switches to control the flow of remaining DC power. This power is used to create sparks responsible for material removal [37].

d) **Tool Holder:** It is a tool holding device which holds the electrode for carrying out the machining operations.

Fig 1.6 Tool Holder

e) **Dielectric System:** It supplies the required amount of dielectric fluid to the cutting zone during machining. Dielectric fluid serves following functions:

 1 Acts as insulator between both surfaces.

 2 It performs the role of coolant and carries away the heat produce during machining operation

3. It removes the material from the cutting zone hence acts as the flushing medium. It is the one of the major and critical function of the dielectric system. Poor flushing causes the dielectric fluid to stagnate and tiny particles are build up in the gap. This will reduce the surface finish and material removal rate

f) **Machining Tank:** In EDM machining tank is the actual tank in which machining operation takes place but in PMEDM machining tank is another tank which is smaller than the machining tank of EDM and is placed inside it. Powder is allowed to mix with the oil present in this small tank rather than mixing with the whole of the dielectric oil.

1.7 Important Process Parameters of PMEDM or EDM

Some of the important process parameters of PMEDM are discussed below:

a) **Discharge Voltage:** It is related to the spark gap and breakdown strength of dielectric. Increase in the discharge voltage increase the gap which improves the flushing conditions and result in higher material removal rate, tool wear rate and surface roughness.

b) **Pulse on Time:** The duration of time (μsec) the current is allow to flow per cycle. Material removal is directly proportional to the amount of energy applied during this on time. This energy is really controlled by the peak current and the length of the pulse on time.

c) **Arc Gap:** It is the distance between the electrode and workpiece during EDM process. It may also be called spark gap. Spark gap can be maintained by servo system.

d) **Pulse off Time:** It is the duration of time between the sparks. This time allows the molten material to solidify and to be wash out of the arc gap. This parameter affects the stability of the arc. If the pulse off time is too short, it will cause sparks to be unstable.

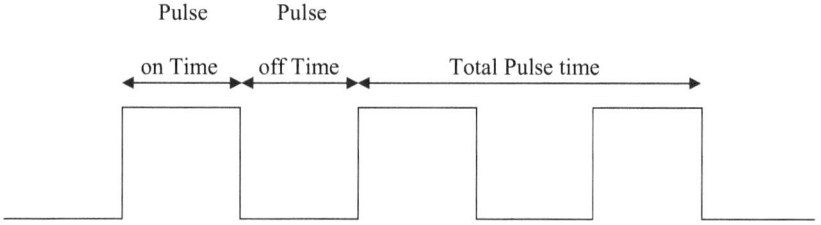

Fig 1.7 Concept of pulse on time and pulse off time [38]

e) **Discharge Current:** It is the amount of power used in electric discharge machining. It is measured in ampere and one of the most important parameter in EDM. High current improves the MRR but at the cost of tool wear and surface finish.

f) **Duty Cycle:** It is a percentage of the on-time relative to the total cycle time. This parameter is calculated by dividing the on-time by the total cycle time (on-time pulse off-time).

$$\text{Duty Cycle} = \frac{T_{on}}{T_{on} + T_{off}}$$

g) **Polarity:** Polarity can be straight or reverse. In straight polarity tool electrode is connected to the negative terminal of power supply and workpiece is connected to positive terminal of power supply and in reverse polarity connection of terminal to workpiece and tool electrode is opposite to that of straight polarity. Straight polarity is normally used polarity in EDM. The negative polarity has high material removal rate and low surface roughness as compared to positive polarity of the tool in EDM [39]. In general, polarity is determined by the experiments and depends on the tool material, work material, current density and pulse length combinations.

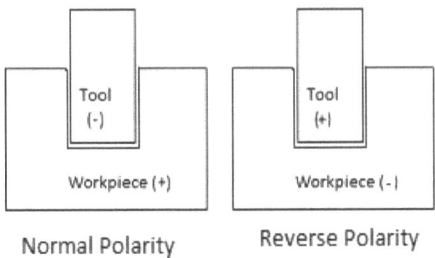

Fig 1.8 Normal and Reverse Polarity

1.8 General Requirements of Dielectric Medium in EDM

In EDM, fluids are used as the dielectric medium. Hydrocarbon oils like kerosene oil, paraffin oil, lubricating oil are widely used as dielectric medium. Deionized water can be used be used as dielectric fluid which gives MRR and TWR. For a good dielectric fluid, it should have the following some characteristics [40]:

a) It should possess high degree of fluidity.

b) It should act as effective cooling medium.

c) It should be able to deionizes the gap immediately after the spark has occurred.

d) It should have high dielectric strength.

e) It should be cheap and easily available.

f) It should be neutral to the workpiece and operator.

1.9 Flushing

Flushing is the most important function in any electrical discharge machining operation. Flushing is the process of introducing clean filtered dielectric fluid into the spark gap. There are a number of flushing methods used to remove the metal particles efficiently.

Method of flushing affects the MRR and TWR to some extent. Flushing can be achieved by the following methods:

a) **Injection flushing:** In this flushing method, the fluid is injected into the spark gap through the hole provided in either workpiece or in electrode

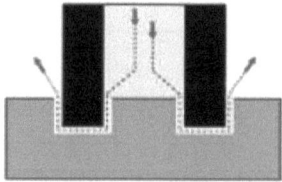

Fig 1.9 Injection flushing through electrode [41]

b) **Side flushing:** In this type flow of dielectric is from one side to another. This type of technique is employed when it is impossible to provide drill the holes in either in workpiece or electrode [37].

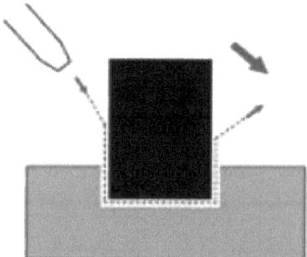

Fig 1.10 Side flushing [41]

c) **Suction flushing:** In this method, the dielectric is sucked either by tool or electrode.

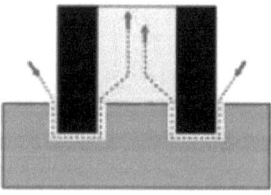

Fig 1.11 Suction flushing through electrode [41]

d) **Flushing by Dielectric Pumping:** Flushing is obtained by using electrode pulsation movement. When the servo control raises the electrode, the gap increases. This results in cleaning of dielectric by suction into mix with contaminated fluid and as electrode is lowered the particles are flushed out.

1.10 Applications of EDM

a) It is used for forging, extrusion, wire drawing and thread cutting.
b) It is used for machining sharp edges and corners that cannot be machined effectively by other machining process.
c) Ceramic materials that are difficult to machine can be machined by the EDM machining process.
d) It is used for drilling curved holes.
e) It is used to machine extremely hard materials that are difficult to machine like alloys, tool steels, tungsten carbides etc.
f) The EDM process is most widely used by the mould-making tool and die industries, but is becoming a common method of making prototype and production parts, especially in the aerospace, automobile and electronics industries in which production quantities are relatively low [40].
g) Higher Tolerance limits can be obtained in EDM machining. Hence areas that require higher surface accuracy use the EDM machining process.

1.11 Advantages of EDM

a) Any material that is electrical conductive can be machined regardless of its hardness, toughness, strength and microstructure etc.
b) Complicated die contours in hard materials can be produced to a high degree of accuracy and surface finish.

c) No stresses are produced in the workpiece as there is no physical contact between the tool and workpiece.
d) Thin fragile sections such as webs or fins can easily machined without deforming the part.
e) X, Y and Z axes movements allow for the programming of complex profiles using simple electrode [38].

1.12 Disadvantages of EDM

a) Only electrical conductive materials can be machined by EDM. Materials like plastics, glass cannot be machined by EDM.
b) Material Removal Rate of EDM is low which limits its application only to machine very hard and difficult to machine materials like carbides.
c) The choice of the electrical parameters of the EDM is largely depends on the material combination of electrode and work piece and EDM manufactures only supply these parameters for a limited amount of material combinations. When machining special alloys, the users have to develop their own technology [40].
d) Lead time is needed to produce specific, consumable electrode shapes.

CHAPTER 2

LITERATURE REVIEW & PROBLEM FORMULATION

2.1 Literature Review

Some literature has been reviewed to find some gaps in EDM process and to find out the effect of different input parameters on MRR, TWR and Surface roughness in EDM process. Some papers are discussed below to get an idea:

Pecas et al., [1] studied the effect of silicon powder concentration on the surface roughness of AISI H13 steel. Peak current, polarity, duty cycle, flushing pressure and powder concentration were taken as input parameters. Taguchi method was used to design the experiment. It was found that surface roughness decreases with increase in the powder concentration of silicon powder and negative polarity of the tool electrode and increases with increase in current, duty cycle and flushing pressure.

Klocke et al., [2] conducted an experiment to study the effect of aluminium powder mixed dielectric on surface roughness of Inconel 718 alloy using tungsten electrode. Polarity, voltage, powder concentration and pulse duration were taken as input parameters. Response Surface methodology was used as the design technique. It was found that powder concentration and pulse duration has the significant effect on the surface roughness. Increase in powder concentration resulted in lowering the surface roughness and vice versa. On the other hand, low surface roughness was found at low pulse duration.

Kansal et al., [3] optimized the process parameters of powder mixed electrical discharge machining (PMEDM) on tool steel using Response surface methodology. Pulse on time, duty cycle, peak current and concentration of the silicon powder added into the dielectric fluid of EDM were chosen as process parameters to study the process performance in terms of material removal rate and surface roughness. The silicon powder suspended in the dielectric fluid of EDM affects both MRR and SR. It was revealed that more improvement in MRR and SR are expected at still higher concentration level of silicon powder.

Yan et al., [4] investigated the influence of the machining characteristics on pure titanium metals using an electrical discharge machining (EDM) with the addition of urea into distilled water. Experimental results indicated that the nitrogen element decomposed from the dielectric that contained urea, migrated to the work piece, forming a TiN hard layer, resulting in good wear resistance of the machined surface after EDM. It was concluded that adding urea into the dielectric, MRR and EWR increased with an increase in peak current. Moreover MRR and EWR declined as the pulse duration increased. The surface roughness deteriorated with an increase in peak current.

Wu et al., [5] studied that addition of aluminium powder in the dielectric can result in obtaining the electric discharge distribution effect. Electrostatic forces among the powder particles make the powder to concentrate at one point in the dielectric. Hence, to separate the powder homogeneously in the dielectric, a surfactant can be used. In the study, effect of adding powder and surfactant in the dielectric on the work piece (SKD steel) after EDM is investigated. With 0.1g/l and 0.25g/l of powder and surfactant respectively, a best distribution effect is found. Positive polarity, gap voltage 90 V, surfactant concentration 0.25g/l, discharge current 0.3 A, pulse duration time 1.5 μs were required for optimum surface roughness. There is 60% improvement in the surface roughness as compared with the surface roughness of the surface machined with no powder in the dielectric.

Kansal et al., [6] performed the experiments on H-11 steel using copper electrode. Peak current, pulse duration, duty cycle and silicon powder concentration were taken as input parameters while the MRR and surface roughness were taken as performance measures.

Taguchi method was used to design the experiments. It was found that pulse duration, powder concentration and peak current affect the MRR and surface roughness significantly.

Dhar and Purohit [7] studied the effect of pulse on time, current and open voltage on the MRR and TWR during EDM with Al–4Cu–6Si alloy–10 wt. % SiC composites. Experiment was performed on PS LEADER ZINC EDM machine using cylindrical type brass electrode of 30 mm diameter. Three level factorial designs were used to analyze the results. In order to establish the relationship among the machining parameters, a second order non linear mathematical model was developed. ANOVA was used to check the significant of the model. It was found that MRR and TWR increase significantly with increases in current.

Kansal et al., [8] studied EDM of AISI D2 die steel by mixing silicon powder in the dielectric fluid of EDM. Copper electrode of 25 mm was used as the tool. Six parameters namely peak current, pulse on time, pulse off time, concentration of powder, gain and flushing pressure were considered as input parameters and MRR as output. Taguchi method was used as the design technique. All the input parameters except flushing pressure have the significant effect on the MRR according to ANOVA analysis. Peak current and powder concentration were most effective parameters in causing material removal. Levels of the parameters that produce optimum MRR were Peak current (16 A), powder concentration (4g/l), pulse on time (100micro sec), pulse off time (15 micro sec) and gain (1mm/sec).

Yeo et al., [9] conducted the experiments on SKD 61 steel using dielectric with and without additive and at low discharge energies of 2.5µJ, 5µJ and 25µJ, and it was observed that a considerable difference in crater morphology is seen between craters in dielectric with and without the powder at low discharge energy of 2.5µJ, 5µJ and 25µJ.

More circular shapes with smaller diameters are produced with powder additive as compared to without powder additive. Craters with the additives are smaller and have more consistent depth than in dielectric without additive. It was reported that dielectric with additive, lowers the amount of discharge flowing between the work piece and the tool electrode and slows down the rate at which these charges flow.

Beri and Kumar [10] performed experimentation on EDM of AISID2 steel in kerosene with copper tungsten (30% Cu and 70% W) electrode (made through Powder Metallurgy technique) and conventional Cu electrode. An L18 orthogonal array of Taguchi methodology was used to identify the effect of process input factors (current, duty cycle and flushing pressure) on the output factors (material removal rate and surface roughness). It was recommended to use conventional Cu electrode for higher MRR and CuW electrode made through PM for low SR.

Chiang [11] studied the influence of discharge current, pulse on time, duty factor and open discharge voltage on MRR, TWR and surface roughness during the EDM of Al_2O_3 +Tic mixed ceramic. Response surface methodology was used to design the experimental design matrix. ANOVA was used to analyze the results obtained. It was found that duty factor and discharge current have the significant effect on the MRR. Discharge current and pulse on time effect significantly both TWR and surface roughness. The value of TWR decreases with increase of the discharge current but increases with an increase of duty factor and discharge voltage. The value of surface roughness increases with an increase of discharge current and open circuit voltage but decrease with an increase of duty cycle.

Lin et al., [12] presented EDM of Tungsten carbide using copper electrode to study the effect of varying machining parameters on MRR, TWR and surface roughness. Furthermore, effects of electric discharge energy on heat affected layer and diameter of machining debris were also determined. Results showed that TWR, MRR and diameter of machined debris increased with increase in density of electric discharge energy. When the amount of electrical discharge energy was set to a high level, serious surface cracks on the machined surface of the cemented tungsten carbides caused by EDM were evident.

Pecas and Henriques [13] showed that addition of powder particles in the dielectric of EDM modifies some process variables and results in creating the condition for higher surface quality. Silicon powder concentration and flushing flow rate was varied and surface morphologic analysis was used to evaluate the process. It was found that addition of silicon powder reduces the crater dimension, white layer thickness and surface roughness however, increases the powder concentration causes the crater dimension to reduce slightly. In order to improve the process polishing capability, it is required to have accurate control of the flushing flow and powder concentration. It was observed that the powder concentration in the range 2 to 3 g/l is necessary to have better surface morphology.

Furutani et al., [14] investigated the influence of discharge current and pulse duration on the titanium carbide deposition process by electric discharge machining using titanium powder mixed dielectric. Copper electrode of 1mm was used to prevent the flushing of working oil from the gap between the electrode and the workpiece. TiC layer deposited on the workpiece as soon as the Ti powder reacted with the cracked carbon from the working oil. A thickness of the TiC layer became the maximum at certain discharge current and pulse duration. The maximum hardness of the deposition was 2000 Hv. The workpiece about 10 μm beneath its surface was also hardened because of the dispersion of TiC.

Prihandana et al., [15] presented a new method that consists of suspending micro-MoS_2 that consists of suspending micro-MoS_2 powder in dielectric fluid and using ultrasonic vibration during μ-EDM processes. Taguchi method was used to optimize process parameters which were concentration of micro-powder, tool electrode material, ultrasonic vibration of the dielectric fluid and workpiece material, to increase the MRR. It was observed that the introduction of MoS_2 micro-powder in dielectric fluid and using ultrasonic vibration significantly increase the MRR and improves the surface quality.

Beri et al., [16] conducted an experiment on ZNC EDM machine to study the effect of aluminium powder mixed dielectric on machining characteristics of Hastealloy. Copper electrode was used as a tool. Concentration and grain size of the powder were taken as input parameters. Addition of powder resulted in lowering TWR and improves surface finish as well as MRR of the Hastealloy. Very small and large size of powder particle resulted in reducing the MRR but increasing the surface roughness of Hastealloy. %WR decreases by using small size particles and vice versa.

Kumar et al., [17] found that significant amount of material transfer takes place from the manganese powder suspended in dielectric fluid to the machined surface under appropriate machining conditions which changes the surface composition and its properties. OHNS die steel was used as the workpiece. They reported that percentage of manganese increased to 0.95% from 0.52% and that of carbon to 1.03% from 0.82% that result in increase in the micro hardness. For surface alloying, favorable machining conditions were found to be low peak current (4 A), shorter pulse on-time (5μs) and longer pulse offtime (85μs).

Sharma et al., [18] used cold treated copper electrode and conventional copper electrode to study the effect of graphite powder mixed EDM on TWR during EDM of Hastelloy. Concentration of powder, polarity, electrode type, peak current, pulse on time, duty cycle and retract distance were taken as input parameters. It was concluded that addition of powder particles in the dielectric and the use of cold treated electrode resulted in reducing tool wear.

Ojha et al., [19] did the parametric optimization for material removal rate (MRR) and tool wear rate by suspending chromium powder in dielectric using EN 8 as work steel. Response surface methodology (RSM) was used to plan and analyze the experiments. Average current, duty cycle, electrode diameter and concentration of chromium powder added into dielectric fluid of EDM were chosen as process parameters to study the PMEDM performance in terms of MRR and TWR. It was observed that current, powder concentration and electrode diameter affect the MRR and TWR significantly. MRR shows the increasing trend for powder concentration. On the other hand, TWR increases with lower range of powder concentration but then decreases.

Anand et al., [20] developed the multi objective model for machining of AISI202 using copper electrode as cutting tool and kerosene oil as the dielectric fluid. Grey relational analysis was used to optimize the machining parameters Discharge current, pulse off time and pulse on time. MRR and surface roughness were taken as performance measures. Grey relational grade is obtained from the grey relational coefficient of MRR and surface roughness. The results indicated that discharge current was the main parameter affecting MRR. Therefore by properly adjusting control factors, work efficiency and product quality can be increased.

Nayak et al., [21] conducted the experiments on EN-31 steel using aluminium mixed dielectric with copper electrode. Response surface methodology was adopted to analyze the effect of powder concentration, pulse on time, duty cycle and peak current on MRR and surface roughness. Non sorted genetic algorithm was used to optimize the responses such that a set of mutually dominant solutions were found over a wide range of machining parameters.

Singh et al., [22] carried out a study on H-13 steel by mixing aluminium powder of particle size 325 μmm in the dielectric. Solid copper electrode of 8mm diameter was used as the tool. Experimental design was based on L18 orthogonal array. Polarity, peak current, pulse on time, duty cycle, gap voltage and powder concentration were input parameters. ANOVA analysis was used to find the significant factors affecting the surface roughness of H13 steel. It was found all the input parameters affect the surface roughness. Higher peak currents produce rougher surface in EDM process. Negative polarity of the tool electrode is desirable of lowering the surface roughness. Also addition of powder in the dielectric lowers the surface roughness.

Sayed and Palaniyandi [23] carried out the EDM experiments on W300 steel using electrolytic copper as the electrode. Peak current, pulse on time, polarity and concentration of aluminium powder were taken as input parameters. MRR, surface roughness and white layer thickness were taken as performance measures. Taguchi method was used to design the experiments. It was found that distilled water mixed with aluminium powder improves the performance of MRR, SR & WLT. High MRR is obtained in positive polarity, whereas better surface quality (surface roughness and white layer thickness) is achieved in negative polarity. Hence for rough machining, positive polarity can be selected to achieve higher MRR and during finishing a better surface is achieved by changing the polarity.

Agarwal and Modi [24] studied the effect of current, pulse duration, wheel speed, duty cycle and Zinc powder concentration on MRR and surface roughness during EDM of AISI D2 steel. Response surface methodology was used as the technique to design the experiment. It was found that the highest MRR is achieved when current, pulse on time and wheel speed are at peak levels. Similarly the highest MRR is achieved when the duty cycle is at the lowest level. Powder mixed EDM resulted in less surface roughness as compared to non powder mixed dielectric.

Aggarwal et al., [25] conducted an experiment on MMC (Al-SiC with 20-25 vol. % SiC) to study the effect of EDM current, pulse on time, pulse off time, and and powder concentration on TWR using ANN (artificial neural network technique). It was concluded that mixing graphite powder in dielectric significantly reduces the TWR during the machining of MMC. The peak current is the most significant control factor affecting TWR, followed by powder concentration. Also the developed ANN model was found to be reliable and adequate to predict the TWR with negligible prediction error.

Mathapath et al., [26] used the ASI D3/ HCHCR steel to study the effect of chromium powder mixed dielectric on MRR of PMEDM. Pulse on time, pulse off time, peak current, tool electrode lift time and powder concentration were taken as input parameters. It was found that MRR increased with increase in powder concentration and peak current but decreased on increasing the pulse off time. Also as the electrode lift time has increased, the MRR also increased.

Vhatkar et al., [27] used fine silicon powder mixed dielectric to investigate the effect of peak current, pulse on time, duty cycle, gap voltage and concentration of powder on MRR and surface roughness. H-11 steel was used as the workpiece. Taguchi method was used to analyze the results. It was found that the MRR increased to a greater extent and the surface roughness showed the decline trend by adding the silicon powder in the dilelctric. Silicon gives better results in terms of MRR and surface roughness.

Syed et al., [28] employed the Response Surface Methodology as the optimization technique to study effect of peak current, pulse on time and powder concentration on the white layer thickness using mild steel as the workpiece. It was found that low thickness of white layer of 17.14 μm is obtained at high concentration of 4g/l and low peak current of 6 ampere with pulse on time of 120 μseconds.

Goyal et al., [29] used the aluminium powder mixed dielectric to study the performance characteristics of AISI 1045 steel. Electrical parameters namely current, voltage, pulse on time and duty cycle remains constant while grain size and concentration of powder were varied. Taguchi method was employed to optimize the results. Grain size and concentration have the significant effect on MRR and surface roughness.

Singh et al., [30] studied the effect of various process parameters namely powder concentration, peak current, pulse off time, tool electrode diameter and flushing pressure of powder mixed EDM (PMEDM) to reveal their impact on material removal rate (MRR) of EN-8 steel by mixing Zinc (Zn) powder to kerosene dielectric. Taguchi's L-27 Orthogonal Array (OA) design is considered to design and analyze the experiments. It was found that powder concentration, peak current and interaction of both have the significant effect on MRR. On the other hand pulse off time and tool diameter was considered as non significant factors for MRR.

Kolli et al., [31] studied the effect of Boron carbide powder mixed into the dielectric fluid during EDM of Titanium alloy. Pulse on time, pulse off time and powder concentration were taken as process parameters and MRR, TWR and surface roughness were taken as performance measures. It was concluded that addition of powder in the dielectric has significantly improved the electrical discharge density and spark gap, which improve the MRR. TWR increases with increase in powder concentration. High concentration levels of boron carbide powder resulted in good surface finish, less crater and cracks.

Long et al., [32] carried out the experiment to study the effect of various titanium powder concentrations on the material removal rate (MRR), tool wear rate (TWR), and surface roughness (SR) using powder mixed electrical discharge machining (PMEDM) was carried out. The workpiece material used was SKD61 hot work steel and the electrode tool material was copper (Cu). The polarity of the electrodes and the concentrations of abrasive powder in the dielectric fluid were used as the input process parameters. Results showed that titanium powder suspended in the dielectric fluid of the EDM process enhanced the MRR without increasing the TWR. TWR and SR were improved in both cases of the polarity.

Kaldhone et al., [33] conducted an experimental study on the effect of powder mixed dielectric on machining performance in EDM of tungsten carbide. MRR was taken as output parameter and peak current, duty factor, pulse on time, peak current, powder type, workpiece material and flushing pressure were taken as process parameters. Taguchi design was used to analyze the experimental results. Workpieces used in this experiment were W20, W30 and W40 and powders were aluminium, graphite and silicon carbide. It was concluded that current, pulse on time, work piece material, powder type and flushing pressure significantly affect MRR. Duty factor has the least effect on MRR. TWR is mainly affected by current and powder. Finally, it was found that SiC powder and current have the great impact on the MRR of tungsten carbide (W 30).

Gudur et al., [34] studied the effect of silicon carbide powder mixed EDM on machining characteristics of SS 316L. Pulse on time, discharge current and powder concentration were taken as input parameters. MMR was taken as the performance measure. Taguchi methodology was used to design the experiments. It was found that MRR and surface finish are improved using SiC Powder mixed EDM as compared to conventional EDM process.

2.2 Problem Formulation:

From the literature survey, it has been found that EDM is a tremendous machining process that can be used for machining of super tough electrical conductive materials like carbides, hot die steels, inconel, nomonic etc as these cannot be machined by other conventional method. The machines also specialize in cutting complex contours or fragile geometries that would be difficult to be produced using conventional cutting methods. Also being non conventional in nature, there is no contact between the tool and the workpiece, hence no force acts on the workpiece. Because of this, there is no deformation of the workpiece and better dimensional accuracy can be achieved. On the other hand, these machines are not utilized at their full potential because of the failure to run the machines at their optimum operating conditions. Furthermore, low material removal rate and surface finish of the EDM limits its use in industry.

Powder mixed discharge machining (PMEDM) emerges as new technology to remove these limitations to some extent. Literature available on PMEDM is very limited. From the available literature, it is concluded that most of the work is done by using Al, Cu, graphite, Zinc, Silicon powder mixed dielectric to improve the surface finish. Very little work has been reported on the reduction of machining time and tool wear for machining of some hard and tough materials like H13 steel using chromium powder mixed dielectric EDM, which is one of the important requirements of modern industry owing to wide applications of H13 steel. Basically, H13 steel is used in extreme load conditions such as hot-work forging, extrusion, manufacturing punching tools, mandrels, mechanical press forging die, plastic mould and die-casting dies, aircraft landing gears, helicopter rotor blades and shafts, dies for blanking, blending and forging, hot extrusion dies for aluminium, cores, injector pins, nozzles for tin and lead die castings etc. Moreover, little research has been conducted for obtaining the optimum levels of machining parameters using chromium powder mixed dielectric that result in best machining quality in machining of difficult to machine materials like H13 steel.

Keeping in view, the applications of H13 steel, it has been selected as work material for machining using chromium powder mixed dielectric EDM. Copper is selected as the tool for carrying out the experiments because of high electrical conductivity and easy availability. From literature, it has been seen that PMEDM parameters like peak current, pulse on time, duty cycle and powder concentration has great effect on MRR, TWR and Ra. So, due to this reason these parameters were chosen for the experiments.

2.3 Objectives of the Study:

The main objectives of this work are mentioned below:

a) Machining of H13 steel which is very difficult to machine by conventional methods because of high toughness and hardness.

b) Experimental determination of the effects of various process parameters namely peak current, pulse on time, duty cycle and powder concentration on MRR, TWR and Ra.

c) To find the best combination of the input parameters for obtaining the optimal values of surface roughness (Ra), material removal rate (MRR) and tool wear rate (TWR).

d) To find the significant input parameters in the present study.

e) To show the relationship between various input parameters and responses by generating the mathematical equations.

CHAPTER 3

OPTIMIZATION TECHNIQUE USED

This chapter provides the brief introduction about the various techniques used for optimization of input parameters in the present study. Response Surface Methodology (RSM) was used to design the experiment in this research work. Response Surface Methodology was implemented by using Minitab 16 software for constructing the experimental design. The optimization of input parameters was done by using Desirability approach. Minitab 16 was used to optimize PMEDM input parameters. A brief description of these methods is highlighted below:

3.1 Response Surface Methodology

Response surface methodology is a collection of mathematical and statistical techniques for empirical model building. By careful design of experiments, the objective is to optimize the response (output variable) which is influenced by several independent variables (input variables). An experiment is a series of tests, called runs, in which changes are made in the input variables in order to identify the reasons for changes in the output response. Suppose there are two input variables having different levels influence a response variable y, the relationship between input and output variable can be shown as

$$y = f(x_1, x_2) + e \qquad (3.1)$$

Where e is the error observed in response y. If we donate the expected response as per following:

$$E(y) = f(x_1, x_2) = \mu \qquad (3.2)$$

then the surface represented by equation (3.2) is called response surface

$$\mu = f(x_1, x_2) \qquad (3.3)$$

In most of the RSM problems the form of relationship between the response and the independent variable is unknown. Thus the first step in RSM is to find out the suitable approximation for the true functional relationship between y and set of independent variables.

The first-order model approximation of the function f is reasonable when f is not too curved in that region and the region is not too big. First-order model is assumed to be an adequate approximation of true surface in a small region. If the response is well modeled by the linear function of the independent variables, then the approximation function is the first order model which is written as

$$y = \beta_0 + \beta_1 x_1 + \beta_2 x_2 \ldots\ldots + \beta_k x_k + e \qquad (3.4)$$

A second-order model is useful in approximating a portion of the true response surface with parabolic curvature. The second-order model includes all the terms in the first-order model, plus all quadratic terms $\beta_{11} x_{1i}^2$ and all cross product terms like $\beta_{13} x_{1i} x_{3j}$. It is usually expressed as

$$y = \beta_0 + \sum_{i=1}^{k} \beta_i x_i + \sum_{i=1}^{k} \beta_{ii} x_i^2 + \ldots\ldots \qquad (3.5)$$

Almost all problems use these models for mathematical equation formation. The method of least square is used to estimate the parameters in approximating polynomials.

In Response Surface Methodology, normally surface plots and contour plots are used for evaluating the effect of various input parameters on the responses.

3.2 Surface Plot

By using Response Surface Methodology, response surface plot can be generated either for a single pair of input variable or for all possible pair of variables. Surface plots show how the response related to the two input variables. These plots are very fine representation of effect of various input variables on the response variables. A surface plot displays the three-dimensional relationship in two dimensions, with the variables on the x- and y-scales, and the response (z) variable represented by a smooth surface.

The surface plot showing the relationship between two independent variables (C1, C2) and response variable C3 is shown below:

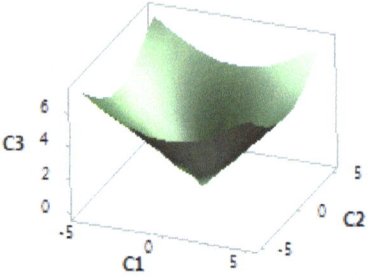

Fig 3.1 Surface plot generated by RSM

3.3 Contour Plot

A contour plot is a graphic representation of the relationships among three numeric variables in two dimensions. Two variables are for X and Y axes, and a third variable Z is for contour levels. The contour levels are plotted as curves; the area between curves can be color coded to indicate interpolated values. Contour plots show how the fitted response relates to two variables. A contour plot provides a two-dimensional view where all points that have the same response are connected to produce contour lines of constant responses. The contour plot showing the relationship between two independent variables (C1, C2) and response variable C3 is shown below:

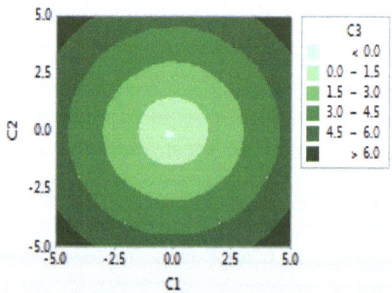

Fig 3.2 Contour plot generated by RSM

3.4 Desirability Approach

For optimization of input parameters of Powder mixed EDM Desirability Approach was used. Minitab 17 software was used to implement the Desirability Approach. In order to get the best set of combination of input parameters for the three responses, Desirability Approach was implemented in the present study.

It is an attractive method for industry for optimization of multiple quality characteristic problems. The method makes use of an objective function, D(X), called the desirability function and transforms an estimated response into a scale free value (d_i) called desirability. The desirable ranges are from zero to one (least to most desirable respectively). The factor settings with maximum total desirability are considered to be the optimal parameter conditions.

The simultaneous objective function is a geometric mean of all transformed responses which is given as:

$$D = (d_1 \times d_2 \times d_3 \times \ldots \ldots \ldots \times d_n)^{1/n} \qquad (3.6)$$

In equation (3.6), n is the number of responses in the measure.

Desirability is an objective function that ranges from zero outside of the limits to one at the goal. The numerical optimization finds a point that maximizes the desirability function. The characteristics of a goal may be altered by adjusting the weight or importance. For several responses and factors, all goals get combined into one desirability function. For simultaneous optimization each response must have a low and high value assigned to each goal. The goal field for responses must be one of the five choices which are:

1) None
2) Minimum
3) Maximum

4) Target

5) In Range

If the goal is none then the response will not be used for optimization. Basically in Desirability method, there are two types of desirability namely individual desirability and composite desirability. Individual desirability evaluates how the settings optimize a single response whereas composite desirability (D) evaluates how the settings optimize a set of responses overall. In the present study, composite desirability will be consider as there are three responses namely MRR, TWR and Ra have to be optimized for different levels of input parameters.

3.5 Optimization Plot by Desirability Method

The optimization plot shows the affect of each factor (columns) on the responses or composite desirability (rows). The vertical red lines on the graph represent the current factor settings. The numbers displayed at the top of a column show the current factor level settings (in red). The horizontal blue lines and numbers represent the responses for the current factor level.

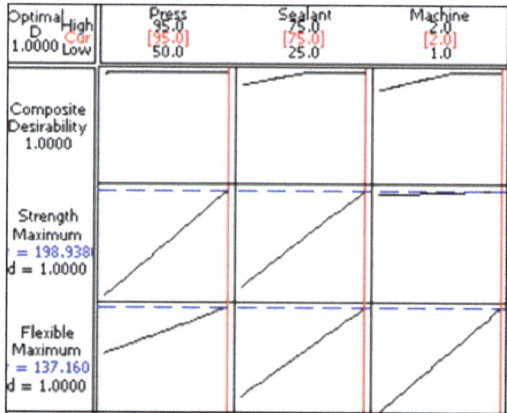

Fig 3.3 Optimization plot by Desirability Method

CHAPTER 4

EXPERIMENTAL SETUP

In this chapter, experimental equipment used for carrying out present study is discussed. For this experimentation, various tools and machines were used to get various observations and results to reach at the desired point.

4.1 EDM Machine

In the present study, the experiments have been conducted on ELEKTRAPLUS PS 35 die sinking Electrical Discharge Machine available at Gaurav Machine Tools Pvt ltd Chandigarh. The pictorial view of the machine is shown in fig 4.1

Fig 4.1 EDM used for experimentation

EDM machine shown in fig 4.1 can vary the peak current from 0.5 to 50 ampere. It has two settings for dielectric fluid i.e 0 and 1. Pulse on time can be varied from 2 to 440 μsec. Pulse off time has the range from 0 to 200 μsec. Spark voltage can be varied from 0 to 15 volt. Fluid Pressure can be varied from 0 to 30 kg/cm². Some other specifications of EDM machine are shown in the following table.

Table 4.1 Specification of EDM machine used for experimentation

Maximum Table Size	330×650 mm
Maximum Workpiece Length	210 mm
Display	LCD
Input Power Supply	3 phase, 415 volt AC, 50 Hz
Connected Load	3 KvA
Dielectric Fluid	Commercial Dielectric Oil
Tank Capacity	140 liltres
Weight	150 kg

4.2 Experimental Setup for Powder Mixed in the Dielectric

For carrying out the experiment by mixing the powder in the dielectric following points should be considered:

a) Entry of the powder to the main tank and filter unit of the machine should be prevented.

b) To prevent the settling of the powder at the bottom of the tank, proper stirring to the powder should be provided.

c) Proper electrical conductivity between the tool and the workpiece should be maintained.

The working tank of the machine is very large having dimensions of 850mm long 550mm wide and 300mm high. Therefore in order to prevent the mixing of powder with the whole dielectric fluid and to avoid the clogging of main filter unit of the machine, a separate tank of 300mm X 150mm X 145 mm is fabricated from the mild steel sheet of 1.5 mm thick and this tank is placed in the working tank of EDM and the machining is performed in this tank. Fixture is placed in the tank to hold the workpiece. Tank is filled with the EDM oil. Small circulation pump is installed in the tank for proper circulation of powder mixed dielectric at the discharge gap. Two permanent magnets are used at the bottom of machine tank to hold the fixture during machining and to separate the debris from the fluid. Set up for powder mixed dielectric EDM is shown in fig 4.2.

Fig 4.2 Experimental set up for powder mixed dielectric EDM

4.3 Preparing Workpiece for Experimentation

A H13 steel plate of dimension 105 mm × 105 mm × 15 mm is taken. As there are 31 experiments have to be performed therefore 31 rectangular pieces are cut from the available plate with the help of power hacksaw. After performing grinding operation for finishing, the rectangular pieces of dimension 28 mm × 20 mm ×10 mm become ready for carrying the experiments. The pictorial view of the final rectangular pieces is shown below

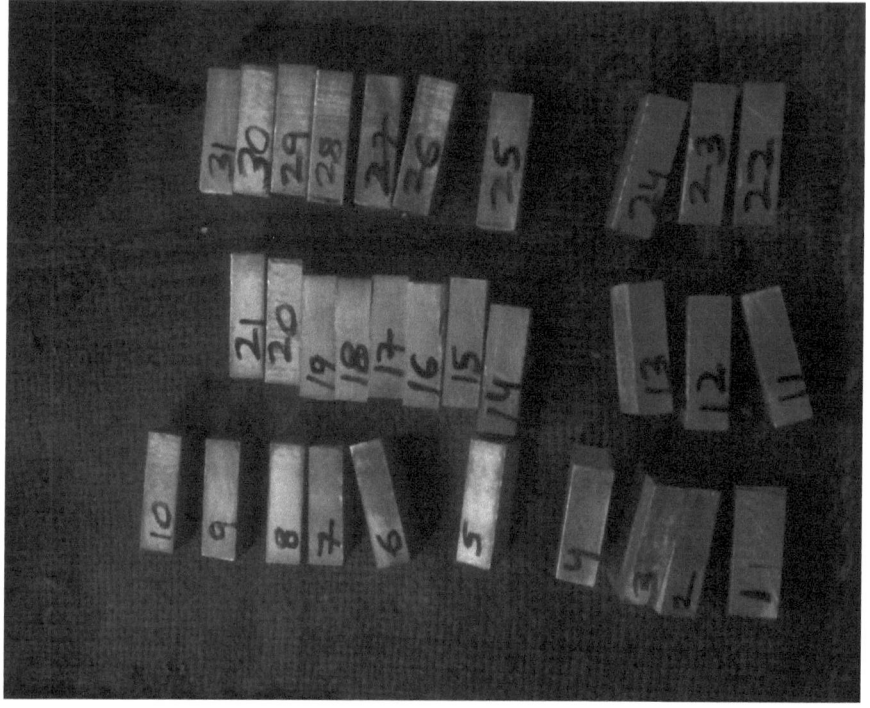

Fig 4.3 Specimens for experimentation

The chemical composition of the H13 steel is shown in table 4.2. This composition is according to the catalogue of the Sehgal die steel company Ludhiana from where material has been bought.

Table 4.2 Composition of H13 die steel

Element	Percentage (%)
Carbon	0.40
Manganese	0.40
Silicon	1.00
Chromium	5.25
Molybdenum	1.35
Vanadium	1.00

4.4 Weighing the Specimen

Before performing the experiments, the weight of all specimens was measured using electronic balance (SONIC) model HBP2000N. The picture of this electronic balance is shown below

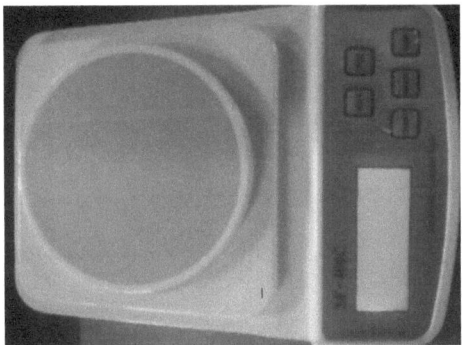

Fig 4.4 Electronic balance

After EDM operation, again the weight of specimen was measured. This was done to calculate the material removal rate (MRR). In the same manner, weight of copper electrode before and after each experiment was measured in order to find out the tool wear rate (TWR).

4.5 Specification of Balance

Some of the specifications of the electronic balance (SONIC) model HBP2000N used in the experimental work are listed in the following table.

Table 4.3 Specification of Electronic balance

Capacity	500 g
Make	SONIC
Readability	0.01g
Repeatability	±0.01 g
Weight	3.8 g Approx.
Data Output	SF400C
Pan Size (Diameter)	140 mm

4.6 Surface Roughness Measurement

Surface roughness is an important output parameter in the present study. Perfilometer, model M4Pi Germany available in the metrology lab of Thapar University Patiala, was used to measure the surface roughness of the specimens. This equipment uses the stylus method of measurement and can measure the surface roughness upto 100 μm. In this study, surface roughness was measured at two different positions with the tracing length of 4 mm and average of the values at these two positions was taken as final reading for analysis the result.

CHAPTER 5

RESULTS & DISCUSSION

This chapter provides the details of the experimental work performed on the EDM machine along with the results of the experimental work.

5.1 Response Surface Methodology

In the present study, Response Surface Methodology (RSM) is used to design the experiments. RSM is implemented with the help of MINITAB 16 software. For designing the experimental matrix, four parameters namely Peak current, Pulse on Time, Duty cycle and Powder concentration are taken as input parameters. Each parameter has three levels in which they are varied. MRR, TWR and Ra are considered as output parameters (desired responses) which will judge the performance of Powder mixed EDM when the values of input parameters will be varied in different levels.

As RSM is used to design the experimental matrix with the help of MINITAB 16 software, therefore there will be total 31 experiments as there are four input parameters (Peak current, Pulse on time, Duty cycle and Powder concentration) are taken in an experiment. Type of design used in the present study is central composite design (CCD). The worksheet of this step is shown below:

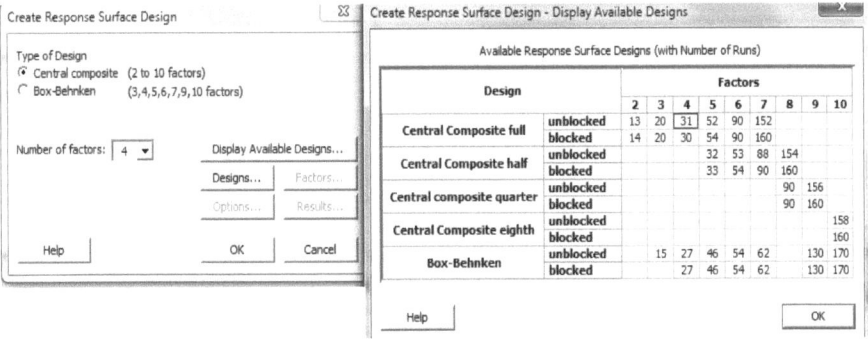

Fig 5.1 Design of experiment by RSM using MINITAB software

5.2 Input Parameters and their Levels

Among the four input parameters, the levels of three input parameters namely Peak current, Pulse on Time and Duty cycle are selected according to the working range of the machine while the levels of the fourth parameter which is powder concentration are selected on the basis of literature review. Input parameters with their levels are listed in the following table:

Table 5.1 Input Parameters with Levels

Parameter	Code	Level 1	Level 2	Level 3
Peak current (Amp)	A	10	15	20
Duty Cycle (%)	B	4	6	8
Pulse On Time (μs)	C	100	150	200
Powder Concentration (g/l)	D	5	10	15

Above mentioned parameters are the parameters which are varied to get the best result of MRR, TWR and Ra. Apart from these parameters, there are some parameters which remain constant throughout the experiment. These parameters along with their values are given in table 5.2.

Table 5.2 Constant Parameters

Work material	H13 die steel
Diameter of copper electrode	16 mm
Fluid pressure	0.5 kg/cm^2
Peak voltage	120 volt DC
Machining time	8 minutes
Gap	3 mm
Pulse off time	180 μseconds
Polarity	Straight

Each run is performed by keeping these above parameters constant at their values given in the table 5.2.

5.3 Responses Variables

In this study, three responses namely MRR, TWR and Ra are selected for evaluating the performance of Powder mixed EDM. Therefore, it is necessary to evaluate the values of MRR, TWR and Ra.

1. **Evaluation of MRR**: The material removal rate is expressed as the ratio of the difference of weight of the workpiece before and after machining to the machining time and density of workpiece.

    $$MRR = (Wjb - Wja) \times 1000/(t \times \rho_,) \text{ mm}^3/\text{min.} \tag{5.1}$$

    Wjb = Weight of the workpiece before machining in grams.

    Wja = Weight of the workpiece after machining in grams.

    $\rho_,$ = Density of H13 steel = 7.80 gm/cm^3.

    t = Machining time = 8 min.

2. **Evaluation of TWR:** TWR is expressed as the ratio of the difference of weight of the tool before and after machining to the machining time and density of the tool material.

TWR= $(Wtb - Wta) \times 1000/(t \times \rho)$ mm³/min. (5.2)

Wtb = Weight of tool before machining in grams.

Wta = Weight of tool after machining in grams.

t= Machining time = 8 min.

ρ = Density of copper = 8.96gm/cm³

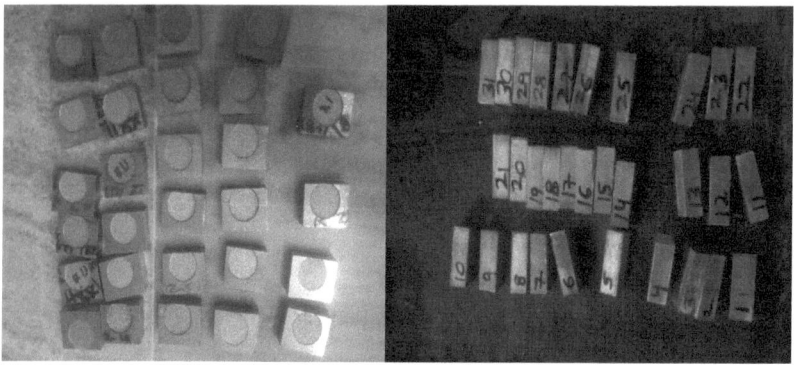

Fig 5.2 (a) Workpiece after machining   **Fig 5.2 (b)** Workpiece before machining

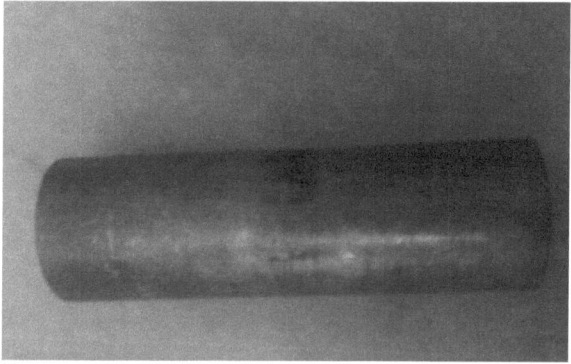

Fig 5.3 Copper electrode used for machining

3. **Evaluation of Surface Roughness (Ra):** In this study, Perfilometer is used to measure the value of surface roughness. Two readings are taken by Perfilometer at two different locations on the surface and the average of these two readings is taken as the final value of surface roughness for this experiment.

5.4 Design Matrix and Observation Table

As there are four input parameters namely peak current, duty cycle, powder concentration and pulse on time therefore total 31 experiments will be carried out according to response surface design. Minitab software is used to design the matrix which is shown in table 5.3

Table 5.3 Design matrix for experimentation

Run	Peak current (Ampere)	Duty cycle (%)	Pulse on Time (μsec)	Powder concentration (g/l)
1	20	4	100	15
2	15	6	150	10
3	20	4	200	15
4	15	6	100	10
5	20	8	200	5
6	15	6	150	10
7	10	4	200	15
8	10	4	100	15
9	15	6	150	10
10	20	8	200	15
11	10	8	200	15
12	20	4	200	5

13	20	6	150	10
14	15	6	150	10
15	20	8	100	15
16	15	6	150	15
17	15	6	150	5
18	10	4	200	5
19	10	6	150	10
20	20	4	100	5
21	10	8	100	15
22	15	6	150	10
23	15	6	150	10
24	15	4	150	10
25	10	4	100	5
26	20	8	100	5
27	15	8	150	10
28	10	8	200	5
29	10	8	100	5
30	15	6	150	10
31	15	6	200	10

According to the above table, the runs are performed on the Electric Discharge machine. Each run is performed for 8 minutes. After EDM operation, the values of all the three responses are calculated for each run which are given in the following observation table 5.4.

Table 5.4 Response values after EDM operation

Experiment No	Weight of workpiece in gram		MRR in mm³/min	Weight of Tool in gram		TWR in mm³/min	Ra in μm
	Wjb	Wja		Wtb	Wta		
1	34.91	34.13	12.5000	87.79	87.67	1.67400	8.31
2	36.33	35.77	8.9743	88.80	88.77	0.41850	8.16
3	35.41	34.51	14.4230	90.24	90.22	0.27900	7.39
4	35.59	35.07	8.3333	88.78	88.77	0.13950	7.56
5	36.27	34.99	20.5128	90.50	90.36	1.95312	8.26
6	35.85	35.32	8.3333	90.28	90.26	0.27900	9.54
7	36.90	36.55	5.6089	90.22	90.20	0.27900	7.51
8	34.76	34.44	5.1282	90.20	90.16	0.55800	6.99
9	35.48	34.90	9.2948	90.27	90.26	0.13950	7.20
10	35.47	33.96	24.1987	88.74	88.72	0.27900	11.70
11	35.42	34.88	8.6598	88.72	88.71	0.13950	8.20
12	36.44	35.77	10.7371	90.96	90.34	0.27900	8.23
13	35.67	34.68	15.8653	87.83	87.80	0.41850	7.66
14	35.36	34.75	9.7756	87.80	87.79	0.13950	7.19
15	36.33	35.10	19.7115	87.68	87.59	1.25558	7.30
16	35.40	34.76	10.2564	87.59	87.56	0.41852	8.50
17	35.90	35.33	9.1346	88.81	88.79	0.27901	8.76
18	35.63	35.42	3.3653	87.85	87.83	0.27901	7.54
19	36.14	35.73	6.5705	87.80	87.79	0.13950	7.34
20	35.49	34.82	10.7371	90.37	90.29	0.13950	8.23

21	35.72	35.13	9.4551	88.71	88.70	1.11607	6.85
22	37.02	36.43	9.4551	88.77	88.75	0.13950	7.40
23	35.33	34.60	11.6987	88.75	88.74	0.27901	9.62
24	36.16	35.72	7.0513	90.25	90.24	0.13950	7.15
25	35.35	35.06	4.7976	88.85	88.84	0.13950	8.95
26	36.25	36.14	17.7884	88.87	88.81	0.13950	8.52
27	36.36	35.60	12.1795	87.81	87.80	0.83750	10.11
28	36.27	35.82	7.2115	87.76	87.75	0.13950	9.22
29	36.91	36.37	8.6538	87.84	87.83	0.13950	7.44
30	32.95	32.35	9.6618	90.26	90.25	0.13950	7.05
31	35.88	35.17	11.3782	90.30	90.28	0.13950	8.14

5.5 Results for Material Removal Rate

For analysis the results of MRR, ANOVA table has been used. This table is generated by the MINITAB 16 software. ANOVA table signifies about the effect of input parameters on the response. It indicates which parameter is significant in a particular process and also provides the information regarding the significance of input parameters on the response variables. It can be seen from ANOVA table that which input parameter affects the response parameter more than the other input parameter. Table 5.5 represents the ANOVA table for material removal rate which is as follow:

Table 5.5 ANOVA table for MRR

Source	DF	Adj SS	Adj MS	F ratio	P value
Model	14	674.118	48.151	56.81	0.0001
Linear	4	618.235	154.559	182.90	0.0001
A	1	428.409	428.409	506.96	0.0001
B	1	166.920	166.920	197.53	0.0001
C	1	5.315	5.315	6.29	0.0231
D	1	17.592	17.592	20.82	0.0001
Square	4	23.036	5.759	6.81	0.0021
A*A	1	6.413	6.413	7.59	0.0141
B*B	1	0.002	0.002	0.00	0.9581
C*C	1	0.144	0.144	0.14	0.7181
D*D	1	0.006	0.006	0.01	0.9321
Interaction	6	32.847	5.474	6.48	0.0011
A*B	1	20.122	20.122	23.81	0.0001
A*C	1	8.312	8.312	9.84	0.0061
A*D	1	1.852	1.852	2.19	0.1581
B*C	1	0.645	0.645	0.76	0.3951
B*D	1	0.057	0.057	0.07	0.798
C*D	1	1.860	1.860	2.20	0.1571
Residual Error	16	13.521	0.845		
Lack of fit	10	6.972	0.697	0.64	0.7471
Pure Error	6	6.549	1.092		
Total	30	687.639			

From the above ANOVA table for MRR, it was noticed that at 95% level of confidence ($p < 0.05$), peak current (A), duty cycle (B), pulse on time (C) and powder concentration (D) all have significant effect on the MRR. The interactions between peak current and duty cycle also interaction between peak current and pulse on time has the significant effect on MRR. The model value F value 56.81 indicates that model is significant. The "Lack of Fit F- value" of 0.64 implies that "Lack of Fit" is not significant. There is 74.71% chance that a "Lack of Fit" this large could occur due to noise. Non significant lack of fit is good as we want the model to fit.

With the help of MINITAB 16 in RSM, a mathematical equation between the MRR and given input factors has been generated as below

$$\text{MRR} = 20.32 - 2.152\,A - 0.31\,B - 0.0832\,C - 0.215\,D + 0.0629\,A*A - 0.008\,B*B$$
$$+ 0.000084\,C*C + 0.0020\,D*D + 0.1121\,A*B + 0.002883\,A*C + 0.01361\,A*D$$
$$+ 0.00201\,B*C - 0.0060\,B*D + 0.001364\,C*D \qquad (5.3)$$

The above equation in terms of coded factors can be used to make the predictions about the MRR for given levels of each factor.

Furthermore, in order to study the effect of individual parameter on the response, main effect plots are generated. When the mean response changes across the levels of a factor, then the main effect occurs. The main effects plot for MRR is shown below in fig 5.4. From the main effects plot it is found that MRR increases almost linearly with increase in current (A). It may be due to the fact that pulse energy increases with increase in current which further increases the rate of heat energy in the discharge channel and hence results in increasing the rate of melting of workpiece at the portion where discharge occurs which enhances the MRR. With increasing the Duty cycle (B), MRR increases continuously because percentage of pulse on time duration relative to total cycle increases which increases the energy density on workpiece and hence the MRR.

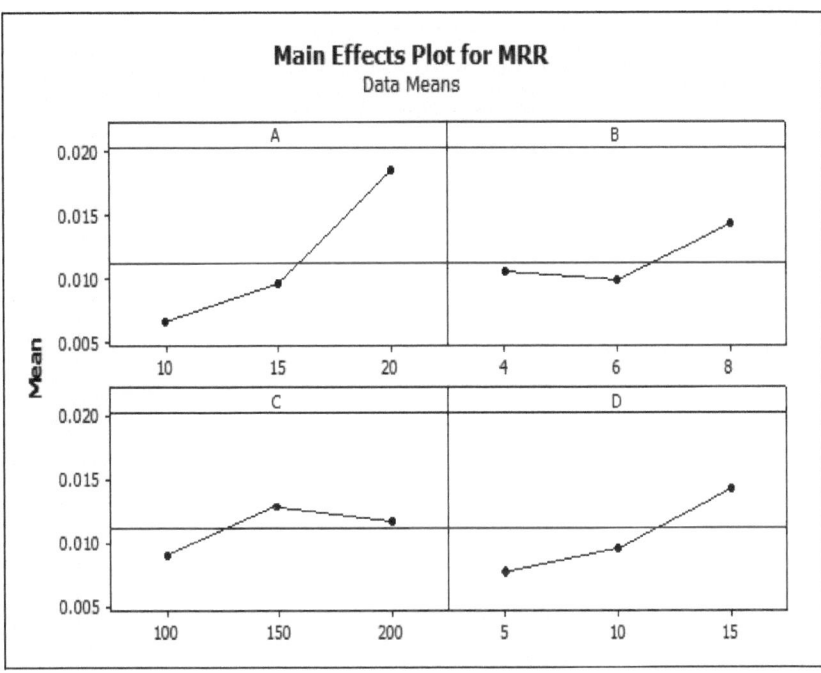

Fig 5.4 Main effects plot for MRR

It can be seen the main effect plot that MRR increases when the value of pulse on time (C) changes from 100 μseconds to 150 μseconds and then decreases as the value of pulse on time increases from 150 μseconds to 200 μseconds. This may be due to the fact that short pulse duration causes less vaporization on the workpiece surface, whereas longer pulse durations make the machining process unstable due to increased possibility of short circuiting. Also increasing the powder concentration (D) in the dielectric increases the MRR as on increasing the powder concentration, the breakdown strength of the dielectric fluid starts decreasing.

In this study, surface plots and contour plots have been also generated by RSM in MINITAB 16. These plots show how the response (MRR) is related to the combined effect of two input parameters in one time.

Surface plots and Contour plots for MRR are shown below:

Surface Plots of MRR

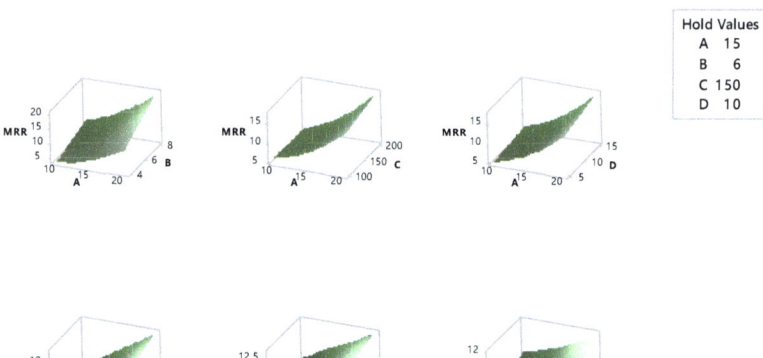

Contour Plots of MRR

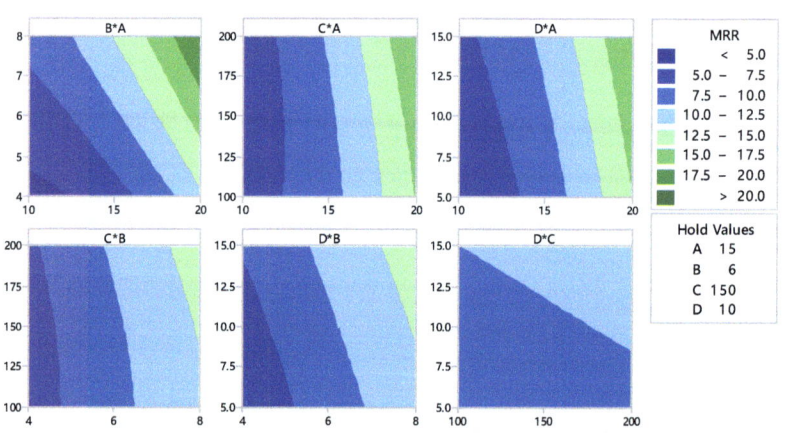

Fig 5.5 Surface plot and Contour plot for MRR

From the surface plots and contour plots for MRR, it is clear that maximum values of all input parameters namely Peak current (A), Duty cycle (B), Pulse on time (C) and Powder concentration (D) result in high MRR. These plots also indicate that maximum material removal rate of more than 20mm^3/min occurs at 20 amp of current and 8% of duty cycle holding the values of pulse on time and powder concentration at 150μsec and 10 g/l respectively.

For determining whether the model meets the assumptions of the analysis, residual plots are plotted for MRR with MINITAB 16 in RSM. These plots indicate whether the residuals are normally distributed, outliers exist in the data. It also indicates whether the variance is constant or nonlinear relationship exists in the data. The residual plots for MRR are shown below in fig 5.6

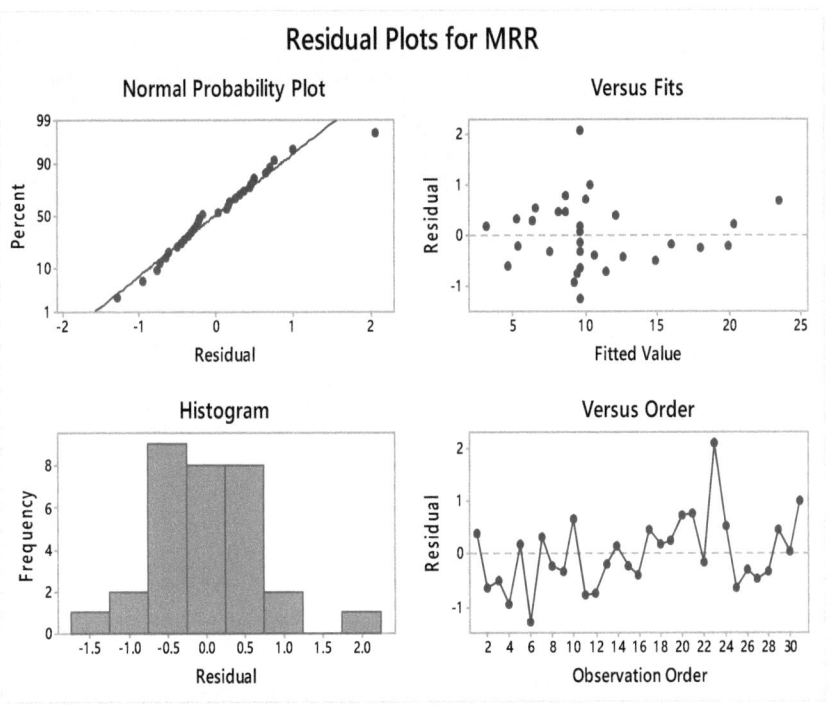

Fig 5.6 Residual Plots for MRR

Normal probability plot indicates that the data are normally distributed as the points make the straight line. As there is no recognizable pattern in the residual versus fitted value plot, therefore nonlinear relationship exists in the data. Histogram indicates that the data are not skewed and variance is constant. Residual versus order of the data indicate that there is no error due to time or data collection order as residuals exhibit no clear pattern.

5.6 Results for Tool Wear Rate

For analysis the results of TWR, ANOVA table has been used. This table is generated by the MINITAB 16 software. Table 5.6 represents the ANOVA table for material removal rate which is as follow:

Table 5.6 ANOVA table for TWR

Source	DF	Adj SS	Adj MS	F ratio	P value
Model	14	5.16232	0.36874	4.41	0.003
Linear	4	2.34238	0.58560	7.00	0.002
A	1	2.09358	2.09358	25.03	0.000
B	1	0.00434	0.00434	0.05	0.823
C	1	0.24337	0.24337	2.91	0.107
D	1	0.00109	0.00109	0.01	0.911
Square	4	1.03425	0.25856	3.09	0.046
A*A	1	0.04785	0.04785	0.57	0.460
B*B	1	0.00004	0.00004	0.00	0.984
C*C	1	0.01132	0.01132	0.14	0.718
D*D	1	0.10965	0.10965	1.31	0.269
Interaction	6	1.78569	0.29762	3.56	0.020
A*B	1	0.17528	0.17528	2.10	0.167

A*C	1	0.23852	0.23852	2.85	0.111
A*D	1	0.07792	0.07792	0.93	0.349
B*C	1	0.39394	0.39394	4.71	0.045
B*D	1	0.31149	0.31149	3.72	0.072
C*D	1	0.58853	0.58853	7.04	0.017
Residual Error	16	1.33841	0.08365		
Lack of fit	10	1.26613	0.12661	10.51	0.665
Pure Error	6	0.07228	0.01205		
Total	30	6.50073			

From the above ANOVA table for TWR, it was noticed that at 95% level of confidence ($p < 0.05$), peak current (A) has significant effect on the TWR. The interactions between a duty cycle and pulse on time also interaction between pulse on time and powder concentration have the significant effect on TWR. The model value F value 4.41 indicates that model is significant. The "Lack of Fit F- value" of 10.51 implies that "Lack of Fit" is not significant. There is 66.5% chance that a "Lack of Fit" this large could occur due to noise. Non significant lack of fit is good as we want the model to fit.

With the help of MINITAB 16 in RSM, a mathematical equation between the TWR and given input factors have been generated as below

$$TWR = 0.93 - 0.056\ A - 0.234\ B - 0.0047\ C + 0.075\ D + 0.00543\ A*A - 0.0009\ B*B$$
$$+ 0.000026\ C*C + 0.00822\ D*D + 0.01047\ A*B - 0.000488\ A*C - 0.00279\ A*D$$
$$+ 0.001569\ B*C - 0.01395\ B*D - 0.000767\ C*D \qquad (5.4)$$

The above equation in terms of coded factors can be used to make the predictions about the TWR for given levels of each factor.

To study the effect of individual parameter on the response, main effect plots are generated. The main effects plot for TWR is shown below in fig 5.7

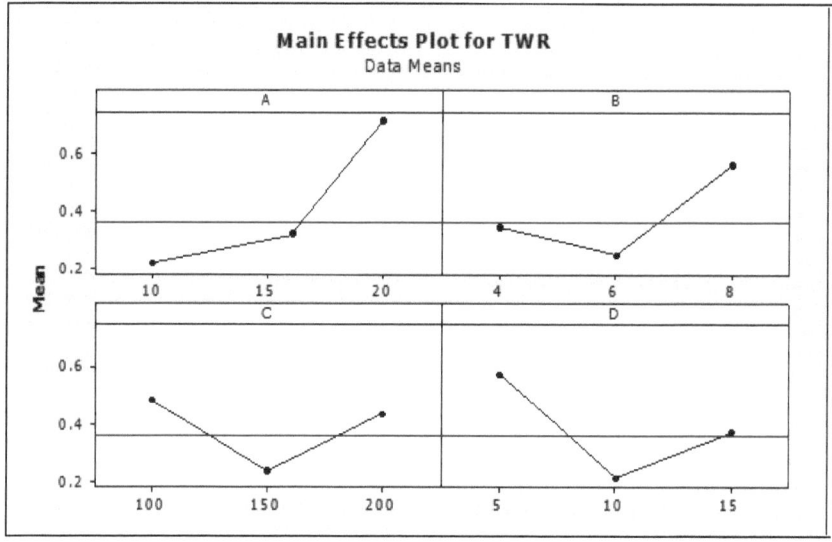

Fig 5.7 Main effects plot for TWR

From the main effects plot it is found that TWR increases with increase in current (A). It may be due to the fact that pulse energy increases with increase in current which further increases the rate of heat energy in the discharge channel and hence results in increasing the rate of melting of electrode at the portion where discharge occurs which enhances the TWR. With increasing the Duty cycle (B) from 4% to 6%, TWR decreases because percentage of pulse on time duration relative to total cycle increases, hence the availability of longer heat transfer time of the molten craters on the tool electrode surface, resulting the lower material removal amount from the molten crater. But as the duty cycle increases to 8% TWR increases because the carbon layer deposited on the tool face may get eroded because of high energy content which enhance the TWR. Increasing the Pulse on time (C) from 100 μsec to 150 μsec, TWR decreases. But on further increasing the pulse on time, TWR increases. It may be due to high energy content at high pulse on time which erodes the carbon layer from the tool face. Overall effect of increasing the concentration (D) of the chromium powder in the dielectric results in reducing the TWR. The reason for this is that the powder particles come in the path of the ions moving towards the electrode surface. It reduces the momentum of the striking ions with the electrode surface and erodes less material from the electrode.

Surface plots and contour plots have been also generated by RSM in MINITAB 16. These plots show how the response (TWR) is related to the combined effect of two input parameters in one time. Surface plots and Contour plots for TWR are shown below:

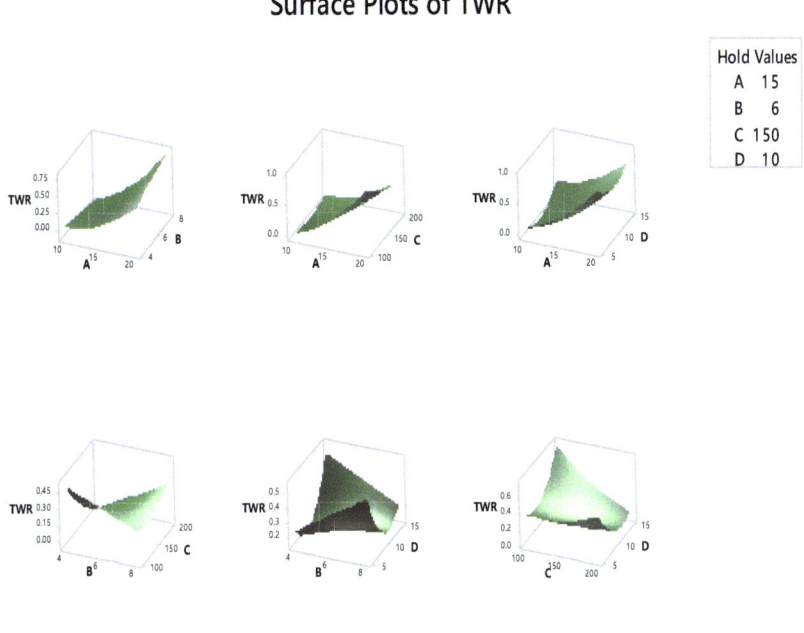

Fig 5.8 Surface plots for TWR

Contour Plots of TWR

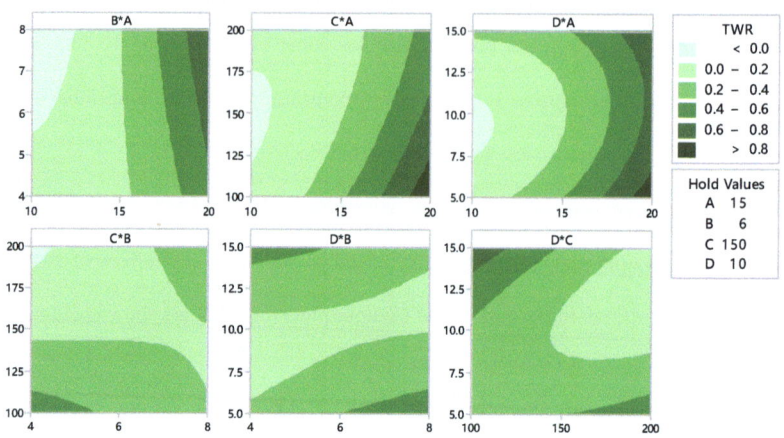

Fig 5.9 Contour plots for TWR

From the surface plot and contour plot of TWR, it is clear that maximum values of all input parameters namely Peak current (A), Duty cycle (B), Pulse on time (C) and Powder concentration (D) result in high TWR which is undesirable. It is clearly understood that TWR will be minimum, when peak current is minimum and pulse on time, duty cycle and powder concentration are at their intermediate values.

For determining whether the model meets the assumptions of the analysis, residual plots are plotted for TWR with MINITAB 16 in RSM. These plots are shown in fig 5.10

Normal probability plot indicates that the data are normally distributed as the points make the straight line. As there is no recognizable pattern in the residual versus fitted value plot, therefore nonlinear relationship exists in the data. Histogram indicates that the data are not skewed and the variance is constant.

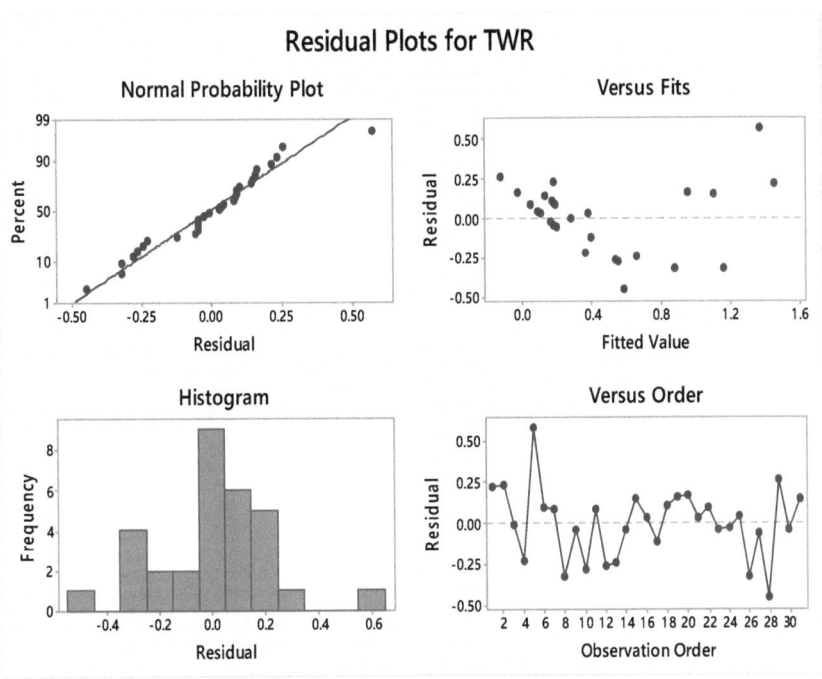

Fig 5.10 Residual plots for TWR

Residual versus order of the data indicate that there is no error due to time or data collection order as residuals exhibit no clear pattern.

5.7 Results for Surface Roughness (Ra)

For analysis the results of Ra, ANOVA table has been used. This table is generated by the MINITAB 17 software. ANOVA table signifies about the effect of input parameters on the response. It indicates which parameter is significant in a particular process and also provides the information regarding the significance of input parameters on the response variable (Ra). Table 5.7 represents the ANOVA table for material removal rate which is as follows:

Table 5.7 ANOVA table for Ra

Source	DF	Adj SS	Adj MS	F ratio	P value
Model	14	18.7510	1.33936	8.92	0.045
Linear	4	7.0247	1.75618	1.21	0.344
A	1	1.7174	1.71742	1.18	0.039
B	1	2.9606	2.96056	2.04	0.172
C	1	2.0268	2.02676	1.40	0.035
D	1	0.3200	0.32000	0.22	0.041
Square	4	1.9658	0.49146	0.34	0.848
A*A	1	1.1532	1.15322	0.80	0.073
B*B	1	0.5573	0.55725	0.38	0.544
C*C	1	0.2601	0.26015	0.18	0.047
D*D	1	0.5573	0.55725	0.38	0.544
Interaction	6	9.7605	1.62675	1.12	0.393
A*B	1	0.5256	0.52562	0.36	0.521
A*C	1	0.0600	0.06003	0.04	0.024
A*D	1	1.6002	1.60023	1.10	0.221
B*C	1	5.1529	5.15290	3.55	0.462
B*D	1	0.7056	0.70560	0.49	0.393
C*D	1	1.7161	1.71610	1.18	0.047
Residual Error	16	23.1964	1.44977		
Lack of fit	10	8.0758	0.80758	0.32	0.946
Pure Error	6	15.1205	2.52009		
Total	30	41.9474			

From the above ANOVA table for Surface roughness (Ra), it was noticed that at 95% level of confidence ($p < 0.05$), peak current (A), pulse on time (C) and powder concentration (D) all have significant effect on the Ra. The interactions between peak current and pulse on time and also between pulse on time and powder concentration have the significant effect on Ra. The model value F value 8.92 indicates that model is significant. The "Lack of Fit F- value" of 0.32 implies that "Lack of Fit" is not significant. There is 94.6% chance that a "Lack of Fit" this large could occur due to noise. Non significant lack of fit is good as we want the model to fit.

With the help of RSM, a mathematical equation between the Ra and given input factors have been generated as below

Ra = 14.92 + 0.590 A - 2.52 B - 0.0061 C - 0.910 D - 0.0267 A*A + 0.116 B*B

- 0.000127 C*C + 0.0185 D*D + 0.0181 A*B + 0.00024 A*C + 0.0127 A*D

+ 0.00568 B*C + 0.0210 B*D + 0.00131 C*D (5.5)

The above equation in terms of coded factors can be used to make the predictions about the Ra for given levels of each factor.

To study the effect of individual parameter on the response, main effect plots are generated. The main effects plot for Ra is shown in fig 5.11

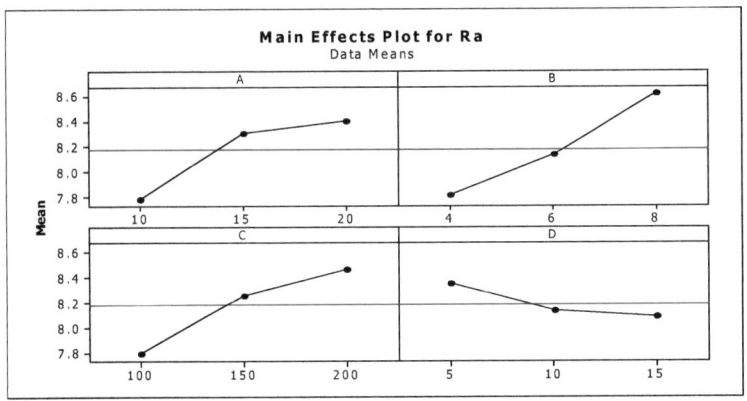

Fig 5.11 Main effects plot for Ra

From the fig 5.11, it is clear that Ra shows an increasing trend with increase in current (A), Duty cycle (B) and Pulse on time (C). It may be due to the fact that high value of current, duty cycle and pulse on time produces larger craters due to large amounts of energy. Hence good surface finish quality can be achieved at low settings of peak current, duty cycle and pulse duration. However increasing the powder concentration in the dielectric decreases the surface roughness. It is due to the reason that powder particles widens the discharge passage which results in easy evacuation of debris from the spark gap and uniform dispersion of discharge energy in all directions causes small and shallow craters on the surface results in decreasing the surface roughness.

Surface plots and Contour plots are generated for Ra using MINITAB 16 software. These plots are shown below

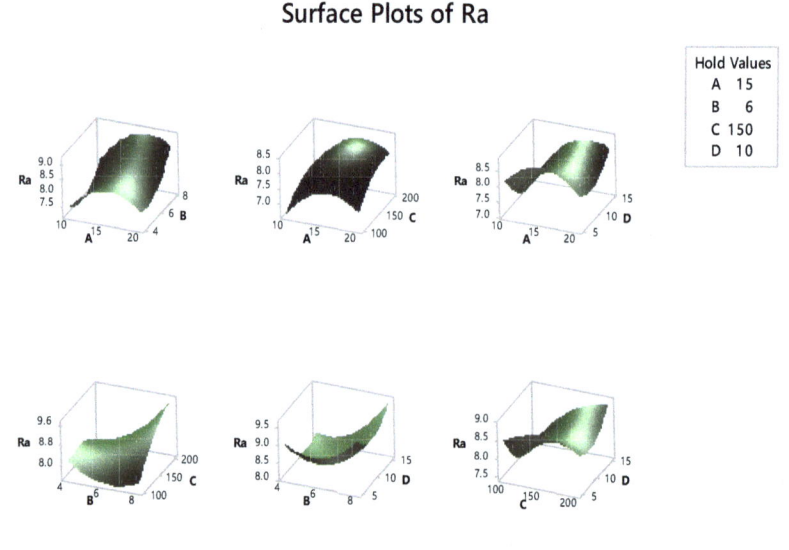

Fig 5.12 Surface plot for Surface roughness (Ra)

Contour Plots of Ra

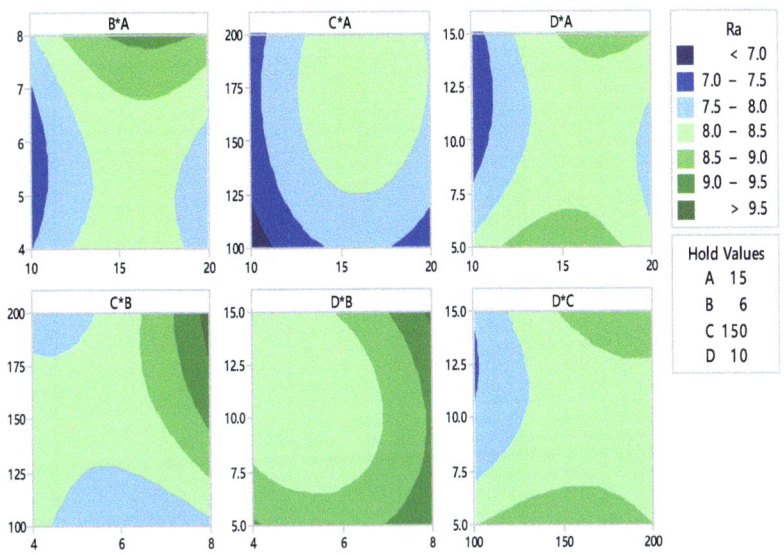

Fig 5.13 Contour plots for Surface Roughness (Ra)

From the contour plots and the surface plots for Ra, it is clear that minimum surface roughness of less than 7 μm is obtained when the peak current is near about 10 ampere, pulse on time is approximately at 120 μsec, powder concentration is at 10 g/l and duty cycle is at 6 %.

For determining whether the model meets the assumptions of the analysis, residual plots are plotted for Surface roughness with MINITAB 16 in RSM. These plots are shown in fig 5.14

Normal probability plot for Ra indicates that the data are normally distributed as the points make the straight line. As there is no recognizable pattern in the residual versus fitted value plot, therefore nonlinear relationship exists in the data.

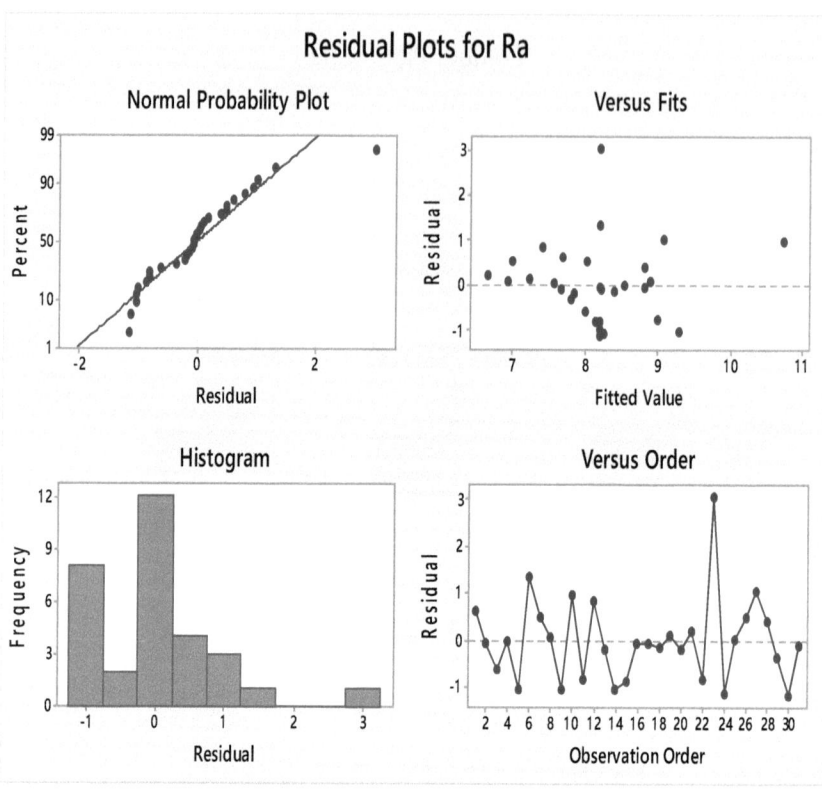

Fig 5.14 Residual plots for Surface roughness (Ra)

Histogram indicates that the data are not skewed and variance is constant. Residual versus order of the data indicate that there is no error due to time or data collection order as residuals exhibit no clear pattern.

5.8 Optimization of the Powder Mixed EDM Parameters

Optimization of the parameters is done by using optimization techniques. In the present study, desirability method of optimization is uses to get the optimal results for the responses. Optimization can be done in following two ways:

1. Single objective optimization.
2. Multi-objective optimization.

5.8.1 Single objective optimization

In single objective optimization, optimization of one desired response at a time is done. Minitab 17 is used to optimize the particular response. All the responses namely MRR, TWR and Ra are treated individually and a set of solutions is generated. After single optimization, combination of input parameters and desired responses are mentioned in the tables below separately for MRR, TWR and Ra.

Table 5.8 Single objective optimization table for MRR

Run	Peak current (A)	Duty cycle (B)	Pulse on Time (C)	Powder concentration (D)	MRR (mm^3/min)	Desirability
1	20	8	200	15	23.5418	0.96847
2	20	8	199.056	14.993	23.3910	0.96352
3	19.995	7.986	200	15	23.3751	0.96128
4	19.941	8	196.339	15	23.200	0.95925
5	20	8	200	14.707	23.295	0.95731

From the table it is clear that maximum MRR predicted by the Desirability method is 23.5418mm^3/min with peak current at 20 ampere, duty cycle at 8%, pulse on time with 15μsec and powder concentration with 15 g/l.

Optimality plot for high MRR with optimum value of parameters is shown below:

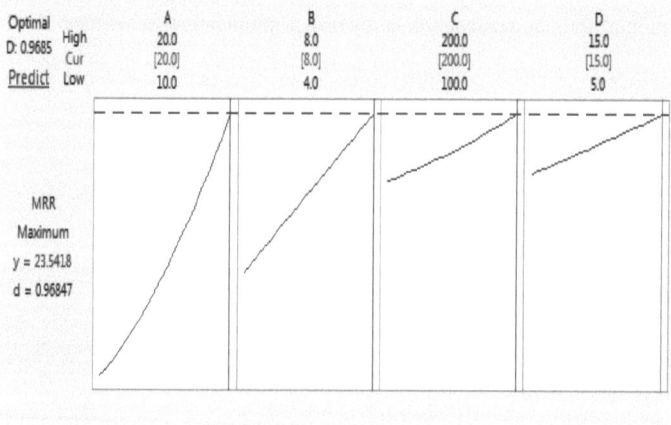

Fig 5.15 Optimal plot for MRR

Single objective optimization table for TWR is given below in table 5.9 with the objective of minimum TWR.

Table 5.9 Single objective optimization table for TWR

Run	Peak current (A)	Duty cycle (B)	Pulse on Time (C)	Powder concentration (D)	TWR (mm^3/min)	Desirability
1	10	8	100	8.6364	0.02530	1.0000
2	10.0120	7.9920	100	8	0.02536	1.0000
3	10.0521	8	99.997	7.9960	0.02552	1.0000
4	11.2091	4	100	5	0.1395	0.9785
5	12.6108	4	99.987	4.992	0.2130	0.9761

From the table it is clear that minimum TWR predicted by the Desirability method is 0.02530 mm^3/min with peak current at 10 ampere, duty cycle at 8%, pulse on time with 100 μsec and powder concentration with 8.6364 g/l with individual desirability of 1.0000.

Optimal plot for with optimum values of parameters with an objective of minimum TWR is shown below:

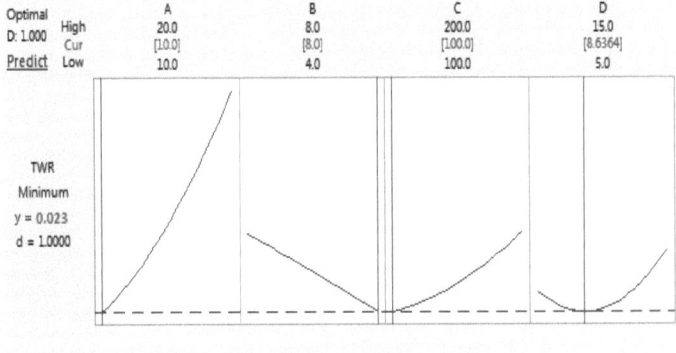

Fig 5.16 Optimal plot for TWR

Single objective optimization table for Ra is given below in table 5.10 with the objective of minimum Ra.

Table 5.10 Single objective optimization table for Ra

Run	Peak current (A)	Duty cycle (B)	Pulse on Time (C)	Powder concentration (D)	Ra (μm)	Desirability
1	10	6.383	100	13.9899	6.3073	1.0000
2	10.212	7.316	100.446	14.448	6.8372	1.0000
3	10.014	7.893	101.907	14.575	6.7867	1.0000
4	10.024	5.773	100.584	14.957	6.8494	1.0000
5	10.062	5.778	100.307	14.965	6.8505	1.0000

From the table it is clear that minimum Ra predicted by the Desirability method is 6.3073μm with peak current at 10 ampere, duty cycle at 6.383%, pulse on time with 100 μsec and powder concentration with 13.9899 g/l with individual desirability of 1.0000.

Optimal plot for with optimum values of parameters with an objective of minimum Ra is shown below:

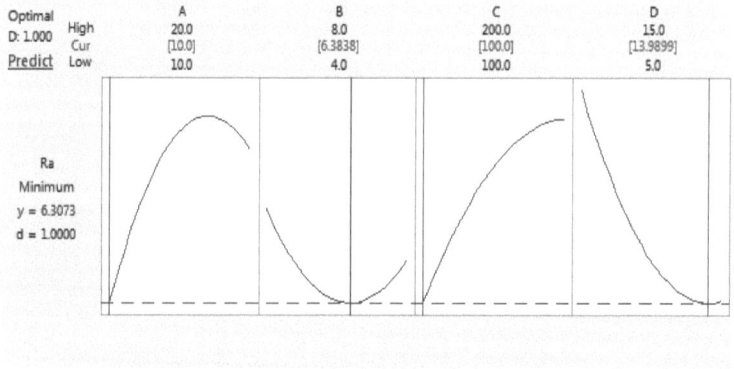

Fig 5.17 Optimal plot for Surface roughness (Ra)

5.8.2 Validity of Desirability Method predicted results

In order to check the validity of the Desirability method predicted results for MRR, TWR and Ra, confirmation test is conducted by carrying out the experiment at mid levels of input parameters and prediction error % is calculated by the following relation

Prediction Error % = (Experimental results- Desirability predicted results) ×100 / Experimental Results

Table 5.11 Validation table for Desirability Method

Run	Experimental Results			Desirability Predicted Results			Prediction Error%		
	MRR	TWR	Ra	MRR	TWR	Ra	MRR	TWR	Ra
1	9.7756	0.13950	8.16	9.6208	0.13623	7.9919	1.58	2.34	2.06

It can be seen from the table that percentage error is within the acceptable limit, therefore results predicted by Desirability Method are valid.

5.8.3 Multi-Objective Optimization

In multi-objective optimization, two or more objectives are optimized at the same time. In machining operations, often there are the cases when improvement in one response may cause the other response to deteriorate. So, it is difficult to find the one single solution to the problem. In such cases, we find the set of solutions to the problem.

Similarly in the present study, there are conflicting responses. Therefore, set of solutions is generated by using Desirability method. There are ten values of desired responses with corresponding values of input parameters which are shown below in the table 5.12

Table 5.12 Multi-objective optimization by Desirability Method

Run	Peak current (A)	Duty cycle (B)	Pulse on Time (C)	Powder concentration (D)	MRR (mm^3/min)	TWR (mm^3/min)	Ra (μm)	Desirability
1	20	7.886	100.146	15	19.7149	1.1144	8.0968	0.7460
2	20	7.891	100.019	14.8210	19.6842	1.1012	8.0603	0.7512
3	20	5.091	200	11.6667	16.1250	0.3161	7.6987	0.7698
4	20	7.843	100	14.252	19.5694	1.0609	7.9583	0.6959
5	20	7.892	100.001	13.8583	19.4887	1.0357	7.8941	0.7148
6	19.787	7.889	100	14.892	19.3326	1.0743	8.0982	0.6509
7	19.802	7.892	100	14.654	19.3101	1.0586	8.0525	0.6572
8	19.728	7.886	100.007	15	19.2482	1.0742	8.1238	0.6483
9	19.997	7.037	178.573	5	17.6258	1.1188	8.5631	0.5884
10	20	6.109	154.778	15	17.8298	0.7482	8.6184	0.6643

From the above table it is clear that run no. 3 gives 16.1250 mm³/min MRR, 0.3161 mm³/min TWR and 7.6987 μm Ra with peak current 20 ampere, duty cycle 5.0909%, pulse on time 200 μsec and powder concentration of 11.667 g/l. This set of solution has the desirability of 0.7698 which is higher among all the desirability of the available set of optimal solutions. Therefore it is the optimal solution for this experiment. Optimization plot for the experiment is given below:

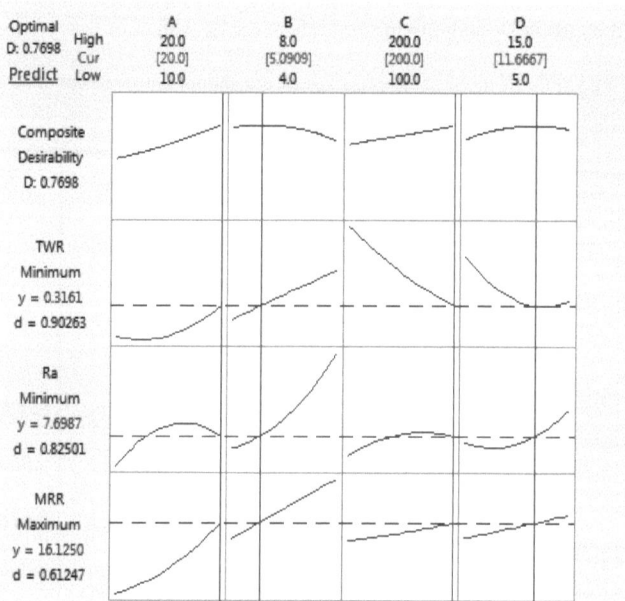

Fig 5.18 Optimization plot for the experiment

CHAPTER 6

CONCLUSION & FUTURE SCOPE

6.1 Conclusion

In the present study, chromium powder mixed dielectric EDM is carried out using copper electrode of 16 mm. Optimization of PMEDM has been done to get the optimal results for MRR, TWR and surface roughness (Ra). The values of four input parameters namely peak current, duty cycle, pulse on time and powder concentration are varied in a particular range to get the best results. Following are the some inferences that came out in the study:

- From the ANOVA tables for MRR, TWR and Ra at 95% of confidence level, it is clear that all the four input factors (peak current, duty cycle, pulse on time and powder concentration) have the significant effect on the MRR. But in case of TWR, it is only the peak current that affects the TWR significantly. For Ra, among the four input parameters pulse on time is most significant followed by peak current and powder concentration respectively.

- From the contour plots and surface plots for MRR, TWR and Ra, it is found that maximum MRR is there when both peak current and duty cycle are increased simultaneously. Minimum TWR is obtained when current is at its low level. In case of surface roughness, minimum value of Ra is obtained when pulse on time and peak current simultaneously are at low levels while powder concentration and duty cycle are at their intermediate levels.

- Residuals plots for MRR, TWR and Ra indicates that the data are normally distributed as the points make the straight line as indicated by normality plot. Also there is no recognizable pattern in the residual versus fitted value plot, therefore nonlinear relationship exists in the data. Histogram indicates that the data are not skewed and variance is constant. Residual versus order of the data indicate that there is no error due to time or data collection order as residuals exhibit no clear pattern.

- Desirability approach is used to optimize the output parameters by using the Minitab 17 software. Two types of optimization gave been done i.e. single objective optimization and multi objective optimization.

- In case of single objective optimization, maximum MRR is 23.5418 mm^3/min with peak current at 20 ampere, duty cycle at 8%, pulse on time with 15μsec and powder concentration with 15 g/l, minimum TWR is 0.02530 mm^3/min with peak current at 10 ampere, duty cycle at 8%, pulse on time with 100 μsec and powder concentration with 8.6364 g/l and minimum Ra is 6.3073 μm with peak current at 10 ampere, duty cycle at 6.383%, pulse on time with 100 μsec and powder concentration with 13.9899 g/l.

- In case of multi objective optimization, ten solutions have been generated by desirability method. Among these solution, the best optimal solution is 16.1250 mm^3/min MRR, 0.3161 mm^3/min TWR and 7.6987 μm Ra with peak current 20 ampere, duty cycle 5.0909%, pulse on time 200 μsec and powder concentration of 11.667 g/l. This set of solution has the desirability of 0.7698 which is higher among all the desirability's of the available set of optimal solutions.

6.2 Future Scope

Some of the future scope of this research is listed below:

- Straight polarity is used in the present work. Effect of reverse polarity on performance measures can be studied in the future work.

- Different type of dielectric powder can be used. In the present study chromium powder has been used. There are several powders like vanadium, nickel, titanium can be used in future practices.
- Circular geometry of the tool is used in the present study. Some other geometries of the tool can be tried in future and results can be compared with the circular tool geometry results.
- Effect of fluid pressure on MRR, TWR and Ra can be studied in future.

REFERENCES

[1] Peças P & Henrique E, "Influence of silicon powder-mixed dielectric on conventional electrical discharge machining". International Journal of Machine Tools & Manufacturing 43(14), 1465–1471, 2003.

[2] Klocke F, Lung D, Antonoglou G & Thomaidis D, "The effects of powder suspended dielectrics on the thermal influenced zone by electro-discharge machining with small discharge energies", International Journal of Material Processing Technology 149, 191–197, 2004.

[3] Kansal H.K, Singh S & Kumar P, "Parametric optimization of powder mixed electrical discharge machining by response surface methodology", Journal of Materials Processing Technology 169, 427–436, 2005.

[4] Biing Hwa Yan, Hsien Chung Tsai & Fuang Yuan Huang S, "The effect in EDM of a dielectric of a urea solution in water on modifying the surface of titanium", International Journal of Machine Tools & Manufacturing 45, 1694–1700, 2005.

[5] Wu KL, Yan BH, Huang FY & Chen SC, "Improvement of surface finish on SKD steel using electro-discharge machining with aluminium and surfactant added dielectric", International Journal of Machine Tools & Manufacturing 45, 1595–1601, 2005.

[6] Kansal H.K, Sehijpal Singh & P. Kumar, "Performance parameter optimization of powder mixed electric discharges machining (PMEDM) by Taguchi method", West Indian journal of Engineering 29(1),81-94, 2006.

[7] Dhar S, Purohit R, Saini N, Sharma A & Kumar G.H, " Mathematical modeling of electric discharge machining of cast Al-4Cu-6Si alloy-10 wt.% SiCp composites", Journal of Materials Processing Technology, 193(1-3), 24-29, 2007.

[8] Kansal, Sehijpal Singh & P Kumar, "Effect of silicon powder mixed EDM on machining rate of AISI D2 Die steel", Journal of manufacturing Processes 9(1), 13-22, 2007.

[9] Yeo S.H & Chiang N, "Comparative study of PMEDM and conventional EDM on surface morphology of SKD 61", Journal of the Chinese society of mechanical engineers, 14 (3), 307-312, 2007.

[10] Beri N & Kumar A, "Performance evaluation of powder metallurgy electrode in electrical discharge machining of AISI D2 steel using Taguchi method", International Journal of Mechanical, Industrial and Aerospace Engineering, 2 (3), 167-171, 2008.

[11] Chiang K.T, "Modeling and analysis of the effects of machining parameters on the performance characteristics in the EDM process of Al2O3+TiC mixed ceramic", The International Journal of Advanced Manufacturing Technology, 37(5-6), 523-533, 2008.

[12] Wang C & Lin Y.C, "Study of electrical discharge machining for Tungsten Carbide", International Journal of Refractory Metals and Hard Materials, 27(5), 872-882, 2008.

[13] Peças P & Henriques, "Effect of the powder concentration and dielectric flow in the surface morphology in electrical discharge machining with powder-mixed dielectric (PMD-EDM)", International Journal Advanced Manufacturing Technology 37, 1120–1132, 2008.

[14] Furutani K., Sato H & Suzuki M, "Influence of electrical conditions on performance of electrical discharge machining with powder suspended in working oil for titanium carbide deposition process", International Journal of Advanced Manufacturing Technology 40 (11), 1093-1101, 2009.

[15] Prihandana G.S, Mahardika M, Hamdi M, Wong Y.S & Mistsui K, "Effect of micro-powder suspensions and ultrasonic vibration of dielectric fluid in micro EDM process- Taguchi approach", International Journal of Machine Tools and Manufacture, 49 (12-13), 1035-1041, 2009.

[16] Beri N, Singh P, Kumar A & Kumar V, "Some experimental investigation on Aluminium powder mixed EDM on machining performance of Hastelloy steel", International Journal of Advanced Engineering Technology, Vol. 1, 28-45, 2010.

[17] Kumar S & Singh R, "Investigating surface properties of OHNS die steel after electrical discharge machining with manganese powder mixed in the dielectric," International Journal of Advanced Manufacturing Technology DOI 10.1007/s00170-010-2536-3, 2010.

[18] Saurabh Sharma, Anil Kumar & Naveen Beri, "Study of tool wear rate during powder mixed EDM of Hastelloy steel", International Journal of Advanced Engineering Technology, E-ISSN 0976-3945, 2011.

[19] Ojha K., Garg R. K. & Singh K. K, "Experimental Investigation and Modeling of PMEDM Process with Chromium Powder Suspended Dielectric", International Journal of Applied Science and Engineering 9(2), 65-81, 2011.

[20] M. Dev Anand & Rajesh R, "The optimization of the EDM process using RSM and Genetic Algorithms", International conference on modeling, optimization and computing, 2012.

[21] Nayak K & Katari M, " Optimization of EN-31 steel by RSM and Genetic Algorithm using Aluminium powder mixed dielectric EDM", International Journal Advanced Manufacturing Technology 39, 1020–1028, 2012.

[22] Singh G, Singh P, Singh B & Tejpal G, " Effect of machining parameters on surface roughness of H13 steel in EDM process using powder mixed fluid",

International journal of Advanced Engineering Research and Studies, Vol. 2, 148-150, 2012.

[23] Syed K. H & Palaniyandi K., "Performance of electrical discharge machining using aluminium powder suspended distilled water", Turkish Journal of Engineering & Environmental Science 36(3), 195 – 207, 2012.

[24] Aggarwal G & Modi M, "Powder- Mixed Electro-Discharge Diamond Surface Grinding Process: Modelling, Comparative Analysis and Multi-Output Optimization using Weighted Principal Components Analysis", Journal of Mechanical Engineering 59, 735-747, 2013

[25] Agrawal A, Dubey A. K, & Shrivastava P. K, "Modeling and Optimization of Tool Wear Rate in Powder Mixed EDM of MMC", 2nd International Conference on Mechanical and Robotics Engineering (ICMRE'2013) Dec. 17-18, Pattaya (Thailand), 1-6, 2013.

[26] Mathapathi U, Jeevraj S, Kumar S & Ramola I.C, "Analysis of material removal rate with powder mixed dielectric", IJCAE 4(3), 316-332, 2013.

[27] Vhatkar, D. R & Jadhav, B. R, "An Experimental Study on Parametric Optimization of High Carbon Steel (EN-31) by using Silicon Powder Mixed Dielectric EDM Process", International Journal of Science and Research 2(1), 431-436, 2013.

[28] Syed, K.H & Kuppan P, "Studies on Recast-layer in EDM using Aluminium Powder Mixed Distilled Water Dielectric Fluid", International Journal of Engineering and Technology 5 (2), 1775-1780, 2013.

[29] Goyal S. & Singh R. K, "Parametric Study of Powder Mixed EDM and Optimization of MRR & Surface Roughness", International Journal of Scientific Engineering and Technology 3 (1), 56-62, 2014.

[30] Singh N, Singh A, Kumar S & Singh G, " Study and analysis of Zinc PMEDM Process parameters on MRR", International Journal of Emerging Technology and Advanced Engineering, Vol. 4, 541-546, 2014.

[31] Kolli M & Kumar A, "Effect of boron carbide powder mixed into dielectric fluid on EDM of titanium Alloy", International conference on advances in Manufacturing and Materials engineering, 115-119, 2014.

[32] Long T, Cuong Ngo & Janmanee P, "Effects of Titanium powder concentrations during EDM machining efficiency of Steel SKD61 using copper electrode", International Journal of Advanced Foundation and Research in science & engineering, Vol. 1, 9-19, 2014.

[33] Kaldhone Y & Kavade M.V, "An experimental study on the effect of powder mixed dielectric on machining performance in EDM of Tungsten carbide", The proceedings of 12th IRF International Conference, 110-115, 2014.

[34] Gudur S & Potdar V, " Effect of silicon carbide powder mixed EDM on machining characteristics of SS 316 material", International Journal of Innovative Research in Science, Engineering and Technology, Vol. 4, 2003-2007, 2015.

[35] www.google.com/ electrical discharge machining images.

[36] Puertas I & Luis C.J, "A study on the machining parameters optimization of electrical discharge machining", Journal of Materials Processing Technology, 143-144, 521-526, 2003.

[37] Benedict G.F, "Non Traditional Machining Process", CRC press, 1987.

[38] Kiranpreet Singh, "Modeling, Analysis, Evaluation, selection and Experimental investigation on EDM process", Thesis submitted to Thapar University, 2013.

[39] Kumar S, Singh R, Singh T.P & Sethi B.L, " Surface modification by EDM- A Review", Journal of Material Processing Technology, 3675-3687, 2009.

[40] Shailesh Kumar, "Experimental Investigation of Machining Parameters for EDM using U shaped electrode of AISI P 20 Tool Steel", Thesis submitted to national institute of technology Rourkela, 2010.

[41] www.google.com/ types of flushing in electric discharge machining images.